T0279096

Praise for *How to Run Wars*

"Ramp up the threat index to severe! Pursue your Orwellian dream of selling war as peace! In this satirical gem, Coyne and Hall skewer the would-be warmongers among us."

—W. J. Astore, Lt. Col., USAF (Ret.), author, *Hindenburg: Icon of German Militarism*

"An essential and timely contribution to the debate on U.S. foreign policy. I smiled in recognition as each ridiculous claim was skewered, and often laughed out loud at the droll, straight-faced accuracy of the account. But then, immediately, and as with all great satire, I was brought up short: actually, this is not funny. The views being skewered here are all real. Without the kind of close scrutiny you'll find in *How to Run Wars*, those destructive and corrosive views will continue to dominate American military doctrine unchallenged."

—Michael C. Munger, professor of political science, Duke University; author, *Is Capitalism Sustainable?*

"*How to Run Wars* works by saying the quiet part out loud. Some readers, even in the liberty movement, might interpret the book literally, without realizing that it is a highly skilled pasquinade, pure lampoonery of the national security state's duplicitous dupes and many minions. It's a Babylon Bee for PhDs, a deeply layered onion of dark humor and satirical insight into the destruction of America by the very people charged with protecting it. Be prepared to laugh, and cry."

—Robert E. Wright, lecturer in economics, Central Michigan University

"Proponents of American hegemony argue that global military interventionism is necessary to protect freedom, liberalism, democracy, and a rules-based international order. But what does it take to run a 'liberal empire'? Economists Christopher J. Coyne and Abigail R. Hall have spent years studying the American national security state. In this book, they explain how to succeed as a member of the national security elite. Doing so requires propaganda, mendacity, repression of dissent, gross violations of civil liberties, willingness to sacrifice human lives, and flagrant violations of international law. In other words, liberal empire entails violating the very liberal principles it purports to protect. If you want to understand the ugly truths about militarism, this book is a great place to start."

> —**Nathan Goodman**, senior fellow, F.A. Hayek Program for Advanced Study in Philosophy, Politics, and Economics, Mercatus Center, George Mason University

"Coyne and Hall eviscerate the national security elites with merciless satiric wit. These elites want to send your children to war, and they need you misinformed and obedient to get away with it. Don't let them. Read this book. Learn their ways. See the truth. Your children's lives depend on it."

> —**Roger Koppl**, professor of finance, Whitman School of Management, Syracuse University

"Chris Coyne and Abigail Hall have done it again, showing how destructive U.S. foreign policy is. But this time they do it by pretending that they are interventionists teaching other interventionists how to make the case for continuous and widespread intervention. It's often humorous, but it's much more. Their serious point is that this is, unfortunately, how many of the intervenors think. Read, laugh, and learn."

> —**David R. Henderson**, professor emeritus of economics, Naval Postgraduate School; research fellow, Hoover Institution, Stanford University

"Behind an insightful analysis of the nation's military activities is a dark warning to Americans: here is what the military-industrial complex is doing to us. *How to Run Wars* explains how we are losing our freedom in the name of freedom."

> —**Randall G. Holcombe**, DeVoe Moore Professor of Economics, Florida State University

INDEPENDENT
INSTITUTE

100 Swan Way, Oakland, California 94621-1428, U.S.A.
Telephone: 510-632-1366 • Facsimile: 510-568-6040 • Email: info@independent.org
www.independent.org

How to Run Wars

How to Run Wars

A Confidential Playbook for the
National Security Elite

Christopher J. Coyne and
Abigail R. Hall

INDEPENDENT
INSTITUTE

Independent Institute
100 Swan Way, Oakland, CA 94621-1428
Telephone: 510-632-1366
Fax: 510-568-6040
Email: info@independent.org
Website: www.independent.org

Cover Design: Denise Tsui
Cover Image: omig/AdobeStock #39908656
Cover Image: schan/123rf.com #18892469

Library of Congress Cataloging-in-Publication Data

Names: Coyne, Christopher J., author. | Hall, Abigail R., author.
Title: How to run wars : a confidential playbook for the national security elite / by Christopher J. Coyne and Abigail R. Hall.
Description: Oakland, CA : Independent Institute, 2025. | Includes bibliographical references and index
Identifiers: LCCN 2024008245 (print) | LCCN 2024008246 (ebook) | ISBN 9781598133929 (cloth) | ISBN 9781598133943 (ebook)
Subjects: LCSH: National security--United States. | Military policy--United States. | Propaganda, American. | Politics and war--United States. | Militarism--United States. | Imperialism.
Classification: LCC UA23 .C6849 2025 (print) | LCC UA23 (ebook) | DDC 355/.033073--dc23/eng/20240314
LC record available at https://lccn.loc.gov/2024008245
LC ebook record available at https://lccn.loc.gov/2024008246

To David Theroux—
our friend, advocate, and ardent supporter of liberty.
Requiescat in pace.

War is the devil's joke on humanity. So let's celebrate Armistice Day by laughing our heads off. Then let us work and pray for peace, when man can break the devil's chains and nations realize their nobler dreams!

—William Allen White, *Forty Years on Main Street*

Contents

Acknowledgments   xvii

PART I   **A Plea to the National Security Elites**   1

   1   A Call to Arms   3

PART II   **The Playbook**   15

   2   Control the Narrative   17

   3   Capture the Media   31

   4   Prepare the Sacrifices   47

   5   Sacrifice Liberty in the Name of Liberty   63

   6   Embrace Top-Down Economic Planning   73

   7   Loosen the Purse Strings   85

   8   Silence Dissent   97

   9   Ignore International Law   111

  10   Do Not Accept Failure   125

PART III **How to Use This Guide** 137

11 Running Wars 139

Notes 149

Bibliography 161

Index 189

About the Authors 197

Acknowledgments

THIS BOOK BUILDS on and extends our earlier co-authored article, "How to Run Wars: A Confidential Playbook for the National Security Elites," *Independent Review* 27, no. 4 (Spring 2023): 612–625. Both projects were initially inspired by Bruce Winton Knight's book *How to Run a War* (New York: Alfred A. Knopf, 1936).

We are grateful to the Independent Institute, and especially Mary Theroux, Christopher Briggs, and William Shughart, for their support of this project. We appreciate the suggestions and comments from three anonymous reviewers. We also thank Anne Lippincott for her editorial work to improve the book.

Chris thanks his colleagues in the F.A. Hayek Program for Advanced Study in Philosophy, Politics, and Economics at the Mercatus Center at George Mason University for their ongoing support and encouragement. He also thanks his family—Rachel, Charlotte, and Cordelia—for their constant love and support.

Abby thanks her colleagues at the University of Tampa for embracing and encouraging her work. She thanks MK Lindsey for her thirty years of friendship, unwavering support, and suggestions for sources for quotes. They were spot-on, as always. She also thanks her husband, Edgar, and her children, Ellie and Will, for their continuous support.

PART I

A Plea to the National Security Elites

I

A Call to Arms

People—both at home and abroad—must be made safe and kept free. As a member of the national security elite, you are one of the few who are capable of achieving these goals.

It is the unquestionable *duty* of the United States—*your* unquestionable duty—not only to uphold but also to spread liberal democratic ideas around the globe. Our doing that not only secures our own interests but also serves as a sort of inoculation against the ideas of our common political and ideological opponents.

The only way that this can be accomplished, however, as you know, is with a proactive, military-forward foreign policy. This policy must be carefully planned, implemented, evaluated, and carried out by those with the required expertise—those within the U.S. national security state.

In other words—by *you*.

Despite words of dire caution about the world and the success of our cause, do not despair, for there is reason for optimism! A good portion of the intellectual class is on your side. In a recent article, political science scholars Michael Beckley and Hal Brands argued that Putin's invasion of Ukraine offers a golden opportunity for the U.S. government to reassert its global dominance. Reestablishing the

U.S.-led international order requires "massive investments in military forces geared for high-intensity combat, sustained diplomacy to enlist and retain allies, and a willingness to confront adversaries and even risk war."[1] Elsewhere, Robert Kagan, a leading advocate for your crucial role in creating world order, argues that Putin's invasion of Ukraine is the result of passivity of past American policymakers who failed to contain Russia. The implication, Kagan argues, is that it "is better for the United States to risk confrontation with belligerent powers when they are in the early stages of ambition and expansion, not after they have already consolidated substantial gains."[2]

Achieving these goals will not be easy, and the required actions are not for the faint of heart. As Stokeley W. Morgan, assistant chief of the Division of Latin American Affairs, stated in 1926, "No facts concerning the United States have been less clearly comprehended both at home and abroad than those which concern her foreign policy."[3] This statement holds true nearly one hundred years later. We, as dutiful U.S. citizens, wish to assist you, the indispensable national security elites, in this critical task.

To that end, we offer this playbook—a manual for carrying out war and foreign intervention in the name of democracy and freedom.

Drawing from our years of studying diverse U.S. military actions —from the U.S. occupation of the Philippines, World Wars I and II, and the wars in Iraq and Afghanistan, to analyses of humanitarian aid, arms sales, the development of military technology, postwar recovery, and domestic extremism and terrorism—we provide the information necessary to carry out successful interventions. In the chapters that follow, we will discuss how to achieve the goals of spreading liberal democracy and stopping the spread of antithetical ideas.

Fortunately, many of the ideas discussed in this book will not be novel. In fact, for many of the topics discussed, the foundations for success have long been laid. You need only to look to history and existing policy to craft plans for success today. Throughout this book, we will use historical examples from U.S. government experiences to illustrate how past members of the national security elite have successfully carried out critical components of foreign policy and intervention.

To be clear, this book is not to be viewed as a retrospective. Quite the contrary—we intend this work to be used by those within the national security state for contemporary and future interventions abroad. With the ongoing wars on drugs and terrorism, the Russo-Ukraine and Israel-Hamas wars, and the myriad of internal and external threats faced today, we have every confidence that this guide will prove useful in your noble efforts to protect the masses at home while spreading democracy and liberty abroad.

Before turning to the playbook, however, we must provide an important caveat: this book is *not* intended for broader public consumption. The reason is simple and one with which you are quite familiar.

The typical voter, put bluntly, is not capable of clearly assessing these questions or policies; they are unable to make the tough decisions required. After years of study—decades between us—we are keenly aware of the complexities of war and the public's untrained and unthoughtful reactions to it. As we will detail in the chapters that follow, a major part of winning wars is the ability to cultivate the narrative members of the public hear, and proactive actions to shield them from information that would distract or otherwise lead them away from supporting your critical and necessary missions.

We have no doubt, however, that this book may fall into the wrong hands and be read by people other than its intended audience, which is you. Do not fret. As we will show, there are ways to mitigate any negative effects this may cause.

Now, on to the task at hand. How should you run wars? How should you go about your business of defending the world and making it safe for freedom—even (especially?) at the point of a gun?

* * *

Here are some preliminary remarks. First, you need to keep firmly in view that the American way of life is in trouble. Democracy and freedom are under attack. We teeter on a knife's edge.

On the one side is liberty, administered by you. On the other is tyranny.

Should you fail to take control and wield your power, the consequences will be dire. President Biden, our current commander in chief, provided you with your charge before the midterm elections in 2022: "Nothing has been guaranteed about democracy in America. Every generation has had to defend it, protect it, preserve it, choose it, for that's what democracy is: It's a choice."[4] Indeed, there are choices to be made.

The reality is even more dire. Neither domestic *nor* global democracy will survive without your firm hand. According to the president's National Security Strategy released in October 2022, there are many threats to this international, inflexible rules-based order. "The most pressing strategic challenge facing our vision," reads the report, "is from powers that layer authoritarian governance with a revisionist foreign policy. It is their behavior that poses a challenge to international peace and stability—especially waging or preparing for wars of aggression, actively undermining the democratic political processes of other countries, leveraging technology and supply chains for coercion and repression, and exporting an illiberal model of international order."[5] The biggest offenders in this regard are clear to anyone with even a cursory understanding of the geopolitical landscape: China and Russia.

According to the U.S. State Department, the Chinese Communist Party (CCP) "poses the central threat of our times, undermining the stability of the world to serve its own hegemonic ambitions."[6] Later, the same report argues that "the PRC [People's Republic of China] … is the only competitor [to the United States] with both the intent to reshape the international order and, increasingly, the economic, diplomatic, military, and technological power to advance that objective."[7]

While this is certainly not the first time that Beijing has attempted to exert undue influence, its boldness demonstrates a new level of arrogance. The Chinese have clearly forgotten the true realities of the world in which they live—one in which the U.S. government is, and must remain, the world's only superpower. The State Department goes on to outline six distinct ways in which the Chinese government exercises unacceptable audacity by threatening the global order and the interests of the United States.

The CCP engages in "predatory economic practices," violating time and again the promises made to the World Trade Organization and its global trading partners. The CCP uses lending and other economic practices to harm other countries and sway opinion on international issues.[8] Yes, we also leverage the power of our banking system and interfere in foreign markets—but, as you know, we do so in favor of democracy, the noblest goal of all. The "Republic" also has engaged in progressive military intervention, particularly within the Indo-Pacific region. Now possessing the world's largest navy, China spends more on its military capabilities than any other nation in the world, with the sole exception of the United States.[9]

That is simply unacceptable. The United States *must* remain the world's premier military force. Yes, we also build bombs and use them—but in the name of advancing freedom. What could be more laudatory than that—or more exculpatory of global harms caused by intervention?

The CCP media, an absolute propaganda machine, has unleashed unparalleled control of information and censorship upon its own populace.[10] While every government must control "disinformation," there is a profound difference between us and the Chinese. We alter and limit information in the interest of freedom, the national good, and the global good. No reasonable person should complain about these efforts.

In addition, China's human rights record is abysmal. From forced sterilization to prisoner abuse, the Chinese government could teach a master class in inhumane treatment. Yes, the U.S. government has its own history of domestic human rights abuses—forced sterilizations included—but any comparison between the two countries demonstrates profound ignorance and must be resisted at all costs.[11] Actions undertaken by the U.S. government have been necessary to promote the common good, U.S. policy, and global peace. The Chinese have no such reasons to justify their actions.

The State Department also notes that, in a time when the world is looking to reduce its environmental impact, China is once again on the wrong side of history. China emits the most greenhouse gases of any nation on the planet.[12] And, as it builds its military, China will only contribute more to the pollution problem. Why? Because military installations, armed conflict, and routine trainings are notorious for their negative environmental impacts.

Now, you might ask, given our own global military boot-print, if this last criticism could also be leveled at the United States. Indeed, it could.[13] But as *the* source of global order, the United States can and must incur this cost. We have a legitimate claim to influence global affairs. The Chinese do not. Mother Earth will forgive her most dutiful child.

It would be incorrect to assume, however, that the Chinese are the exclusive danger. We cannot discuss contemporary threats to the global hegemony of freedom-loving countries—led, of course, by you—without discussing Russia. According to the White House, "Russia poses an immediate threat to the free and open international system, recklessly flouting the basic laws of the international order."[14] For example, we cannot let the world forget Russia's intervention and involvement in the Syrian civil war.[15] The United Nations Office of the High Commissioner for Human Rights estimates that the Syrian conflict alone has killed more than 306,000 people.[16] Our critics might point out our own government's involvement in the Syrian civil war as contributing to the conflict. They might also bring up the fact that more than 350,000 civilians were killed by all parties as part of the U.S. government's post-9/11 wars.[17] But what the critics neglect to mention is that the United States had a noble intention in killing these people, namely, to make the world safe from terrorists. The deaths of civilians as a result of U.S. policy are regrettable but sadly necessary.

And of course Russia badly interfered with our democracy project when, in 2014, the Kremlin annexed Crimea. If there was any doubt about what means Vladimir Putin would employ to achieve his goals, such doubt was removed in 2022 when the Russian military launched a full-scale invasion of the sovereign nation of Ukraine, shattering peace in the region and displacing an estimated 7.5 million people as of October 2022.[18] The Biden White House summarizes the issue succinctly: "Russia now poses an immediate and persistent threat to international peace and stability."[19] As the United States is preparing to send additional materiel to the Ukrainians, it is clear that we, *once again*, will have to clean up a purely European mess. In truth, we could not have dreamed up a better scenario than a provoked Russia now on the march to justify even more U.S. military expenditure and the exploitation of any occasion to use our new, more powerful weapons.

Apart from China and Russia, we have additional foes. Adversaries are always useful to our cause, and the more varied and scattered around the planet they are, the better.

Take Iran, for example. A perpetually antagonized Iran "undermines Middle East stability by supporting terrorist groups and military proxies, employing its own military forces, engaging in military provocations, and conducting malicious cyber and information operations."[20] Continuing with a tradition of supporting terrorism and other malicious activities, a member of Iran's Islamic Revolutionary Guard Corps was charged in August 2022 in a foiled plot to assassinate former national security advisor John Bolton.[21] Former secretary of state Mike Pompeo was also reportedly a potential target.[22] Iran has continued to send a steady supply of weapons to Russia in its unprovoked war against Ukraine, including unmanned aerial vehicles (UAVs) and surface-to-surface ballistic missiles.[23]

Such things just make our existence—*your* existence—all the more necessary.

But the trouble doesn't stop there. Iran has armed and financially supported Islamic extremist groups throughout the Middle East, including Hamas and Palestine Islamic Jihad (PIJ). Though the connection between Iran and Hamas's coordinated attack on Israeli civilians in October 2023 is unclear, the persistent arming of Hamas and similar groups shows an obvious disregard for and a desire to upend the region's geopolitical stability.[24] These things prove that we are more important, more needed, than ever.

Also useful to us? North Korea, which, "while not a rival on the same scale as the PRC and Russia, nonetheless also presents deterrence dilemmas for the United States and its Allies and partners."[25] As an antagonized P'yŏngyang attempts, like Iran, to expand its nuclear capabilities, North Korean dictator Kim Jong Un continues the tradition, established by his father and grandfather, of threatening the United States. The Korean Central News Agency recently reported that the dictator said, "Our party and our government will resolutely react to nukes with nuclear weapons and to total confrontation with all-out confrontation."[26] Speaking on pressures to reduce or forfeit North Korea's weapons for the sake of peace, Kim stated in September 2022 that "there will never be any declaration of 'giving up our nukes' or 'denuclearization,' nor any kind of negotiations or bargaining. … As long as nuclear weapons exist on Earth … our road towards strengthening nuclear power won't stop."[27]

Those who consider North Korea to be merely an irritant should think again. In November 2022, the Hermit Kingdom fired over a dozen missiles toward the South Korean capital of Seoul. One of these missiles landed in South Korean territory for the first time since the ceasefire in the Korean War.[28] Guess who, as a result, wants the U.S. democratic hegemony to be permanent and as expansive as possible? Right, South Korea! And does anybody think Japan would complain? Or Taiwan?

The specific threats posed by these rogue states are in addition to the continued danger posed by terrorism, both globally and within the United States. Though the attacks on the World Trade Center and Pentagon in September 2001 ushered in a new era of intensive and largely, in our opinion, successful, counterterrorism efforts, terror must *always* be spoken of as an imminent threat. For example, the National Terrorism Advisory System Bulletin issued by the U.S. Department of Homeland Security in November 2022 very usefully summarized things this way:

> The United States remains in a heightened threat environment. Lone offenders and small groups motivated by a range of ideological beliefs and/or personal grievances continue to pose a persistent and lethal threat to our Homeland. … Threat actors have recently mobilized to violence, citing factors such as reactions to current events and adherence to violent extremist ideologies. … Targets of potential violence include public gatherings, faith-based institutions, the LGBTQI+ community, schools, racial and religious minorities, government facilities and personnel, U.S. critical infrastructure, the media, and perceived ideological opponents.[29]

In sum, since its founding, the U.S. government has been a domestic and global force for good in the world—despite (or perhaps because of) the bloodshed. Our Founding Fathers clearly saw the potential for the United States to transform the world through adopting and spreading liberal democratic ideals. Thomas Jefferson, for instance, envisioned an "empire for liberty," expanding over the North American continent. Writing to his successor, James Madison,

Jefferson stated that, after planting a U.S. flag on Cuban soil, "we should then only have to include the North [Canada] in our confederacy … and we should have such an empire for liberty as she has never surveyed since creation."[30]

Recognizing the country's privileged geopolitical position, U.S. government leaders from Jefferson onward have worked tirelessly with you and your colleagues, past and present, to spread democracy, liberty, and peace at home and abroad. These efforts have not always gone according to plan, of course. The withdrawal from Afghanistan following a twenty-year effort at establishing peace and stability ended regrettably, with the country's quick collapse into the hands of the Taliban. The U.S. government's war and occupation of Iraq similarly did not end in the way we had intended.

But no matter.

To be sure, these most recent outcomes have caused some to question the capabilities of the national security elite and the advisability of using the U.S. government's geopolitical position, military dominance, and ability to spread liberal ideals. Naive "peacemongers," antigovernment extremists, isolationists, and—if you can believe it—even some military veterans have attempted to persuade the public to support less U.S. government involvement abroad.

Take Jeannette Rankin, the first woman to hold federal office in the United States and a lifelong pacifist, who said, "You can no more win a war than you can win an earthquake. War is the slaughter of human beings, temporarily disregarded as enemies, on as large a scale as possible."[31] Noam Chomsky, a well-known public intellectual and longtime antiwar radical, has made similar claims for decades. "My own concern is primarily the terror and violence carried out by my own state," he writes. "[The United States] happens to be the larger component of international violence."[32]

Though these quotes from Rankin and Chomsky are separated by a century, they're equally wrong. Wars *can* be won. The idea that the U.S. government is some great perpetrator of mass violence across the globe is equally preposterous. The United States has, consistently and persistently, spread liberal ideals around the world. It is a force, literally a *force,* for peace, and you, as members of the national security elite, captain this indispensable, freedom-spreading, *armed* ship of state. Withdrawing from this path now would be disastrous. Our very ideals are at stake.

It is, therefore, both the United States' interest as a country and its God-given moral duty to prevent any kind of withdrawal from taking place. As President Obama once said, invoking Thomas Hobbes's argument about the importance of a strong sovereign, "I have a recognition that us [the United States] serving as the Leviathan clamps down and tames some of these [baser] impulses."[33] All "liberal" empires are illiberal in practice; but some Leviathans are better than others. And is there any doubt that ours is the best in human history?

The United States has found itself in such a position before. As President Eisenhower stated in 1954 regarding the importance of Indochina (former French colonies in Southeast Asia) in the free world,

> you have broader considerations that might follow what you would call the "falling domino" principle. You have a row of dominoes set up, you knock over the first one, and what will happen to the last one is the certainty that it will go over. ... So you could have a beginning of a disintegration that would have the most profound influences.[34]

We face a similar threat today—and the dominoes are already falling. The members of the national security elite are the only ones capable of preventing more from falling and, with effort, of nation-building those that have already fallen.

It is for you, and for the success of these ambitious projects, that we have written this playbook.

PART II

The Playbook

2

Control the Narrative

WINSTON CHURCHILL ONCE noted that "in wartime, truth is so precious that she should always be attended by a bodyguard of lies."[1]

Truer words have never been spoken.

You must control the narrative. For any war or intervention to be successful, you must have the support of the broader populace. Without popular support for U.S. government operations abroad, your hopes of a freer, more prosperous world are smothered in their cradle. Controlling the way policies are framed and disseminated is more important now than ever before due to the widespread availability of information. The public's access to television, the internet, and social media presents a real challenge for your contemporary foreign policy ventures.

We are not the only ones to recognize this issue. In 2018, the Joint Chiefs of Staff issued a Joint Concept for Operating in the Information Environment (JCOIE), discussing the role of information in war: "JCOIE recognizes that individuals and groups today have access to more information than entire governments once possessed."[2] Discussing the challenges facing the military, they ask, "How will the Joint Force change or maintain perceptions, attitudes, and other elements that drive the desired behaviors of relevant actors in an increasingly pervasive and connected [informational environment] to produce enduring strategic outcomes?"[3] A critical question indeed.

How *do* we change, maintain, or otherwise control perceptions and attitudes to generate the desired behavior from the American public? In an ideal scenario, members of the public would simply accept what their leaders tell them—complying with requests, restrictions, and mandates as their patriotic duty.

Unfortunately, reality often deviates from the ideal.

Long gone are the days when the American people rightly understood not only their duty to their government but also their duty to the world. President Kennedy, in his 1961 inaugural address, told the American people, "My fellow Americans: ask not what your country can do for you—ask what you can do for your country. My fellow citizens of the world: ask not what Americans will do for you, but what together we can do for the freedom of man."[4] But the Americans of Kennedy's time were not the Americans of today, and the global citizens of today are not the global citizens of the 1960s. Dissension and criticism are ubiquitous, and ordinary people have forgotten their proper roles in society. Even when the majority support a policy, *someone* is bound to complain—loudly. Pacifists, the antiwar contingent, and some subset of the "intellectual" class are bound to denounce military intervention and call for alternative forms of engagement.

These people are naive fools. Our enemies could be at their front doors with knives and guns, and these simpletons would invite them in for a tête-à-tête.

While *we* know these dissenters lack intimate knowledge of the threats we face, ordinary citizens are not so keenly aware of the dissenters' ignorance. While *we* know that the "olive branch" diplomacy advocated by those opposed to U.S. foreign policy is an ineffective tactic, others are not so discerning and can be persuaded of the tactic's value or at least led to doubt the necessary policy prescription. So, we must be prepared. Thankfully, the U.S. government has a long history of controlling information as it relates to U.S. foreign policy efforts and the realities of the world.

The Committee on Public Information (CPI), otherwise known as the Creel Committee, is an excellent example of what can be done. The purpose of the CPI, created in 1917, was to proactively control the narrative surrounding the ongoing war in Europe to cultivate and maintain broad public support for U.S. involvement in World War I.[5] Everything from the rationing of goods and buying war bonds to victory gardens and conscription was carefully framed to encourage support for the government's activities.[6]

Newspapers, movies, television, and radio all fell under the purview of the CPI. Four Minute Men, patriotic volunteers, were tasked with delivering brief talks (with the information provided by the CPI) at various public gatherings, including meetings at clubs, schools, and churches.[7] The CPI also created its own newspaper, the *Official Bulletin*, which reached a total circulation of 118,000 with daily issues. For a time, schools throughout the nation received copies of the *National School Service* so that the youth of America could also receive correct information about the war.[8]

The need for such an organization was obvious. Writing in 1914, President Wilson highlighted the problems that can occur when citizens have too much information—or information that isn't properly conveyed:

> We have several times considered the possibility of having a publicity bureau. ... The real trouble is that the newspapers get real facts but do not find them suitable to their tastes and do not use them as given them, and in some of the newspaper offices news is deliberately invented. Since I came here I have wondered how it ever happened that the public got the right impression regarding public affairs, particularly foreign affairs.[9]

The CPI was disbanded after the war concluded. But when it was again critical that the public receive the right information on U.S. foreign policy, a new organization would take its place. During World War II, the CPI was replaced by several organizations, including the Office of War Information (OWI) and the Writers' War Board (WWB).

The OWI, created by executive order in June 1942, was intended to "formulate and carry out, through the use of press, radio, motion picture, and other facilities, information programs designed to facilitate the development of an information and intelligent understanding, at home and abroad, of the status and progress of the war effort and of the war policies, activities, and aims of Government."[10] The OWI would have almost complete control over the dissemination of war-related information, determining what messages the American and foreign publics received and when they would receive them.

The WWB, established in December 1941, is a wonderful example of what can be accomplished when private citizens learn to fulfill their patriotic duties to the nation. The WWB was privately operated and did not have a formal government budget. But even though it was technically "independent," the WWB operated through the OWI and worked "as a liaison between American writers and U.S. government agencies seeking written work that [would] directly or indirectly help win the war."[11] According to one official, "[with the WWB] we have the services of almost 5,000 writers; we reach thousands of newspapers; more than 600 radio stations, and have a vast army of writers ready to cooperate in the government's war work."[12]

In today's information age, controlling the foreign policy narrative is of critical importance. If President Wilson thought that newspapers took liberty with the information they offered the American public, he'd be aghast at what constitutes "news" on social media giants such as Facebook and X (formerly Twitter). On these platforms, people can largely say what they wish with little regard for facts or how what they say affects the greater good. And this information spreads faster than anyone previously had ever thought possible.

"They're killing people," President Biden said of Facebook and misinformation about the COVID-19 pandemic.[13] While he was discussing information on public health, make no mistake: misinformation about war and foreign policy is just as dangerous. Even if the lives and liberties lost are not readily apparent, information that leads people away from supporting U.S. security policy is lethal and must be controlled.

Officials have very recently sought to take back appropriate control of the narrative surrounding matters of national security. In April 2022, the Department of Homeland Security (DHS) announced the creation of the Disinformation Governance Board (DGB) to combat misinformation that threatens national security. "The Department [the DHS and the Board] is focused on disinformation that threatens the security of the American people, including disinformation spread by foreign states such as Russia, China, and Iran," reads a DHS report from May 2022.[14] "Such malicious actors often spread disinformation to exploit vulnerable individuals and the American public."[15] The Board had planned to coordinate with large tech companies in an "analytical exchange" that would have allowed for the collection of data and a better understanding of the type of data being disseminated.[16]

But the plan was not to be. As mentioned earlier, naysayers came from every direction to thwart the plans of the Board—calling it "Orwellian" and a gross violation of privacy. The Board was "paused" and subsequently disbanded.

Let us turn this failure into future success, however, by learning two important lessons. First, you must make certain to publicly emphasize a commitment to truth, transparency, and accountability, even though in practice you must make every effort to conceal your work in this area. The DGB was attacked from multiple angles as soon as people learned of its existence. These attacks came not just from uninformed citizens but from elected officials as well.

For example, Jason S. Miyares, attorney general for the Commonwealth of Virginia, wrote to Secretary of Homeland Security Alejandro Mayorkas, "You seem to have misunderstood George Orwell: the 'Ministry of Truth' described in *1984* was intended as a warning against the dangers of socialism, not as a model of government agency."[17] Missouri senator Josh Hawley published "whistleblower" documents related to the DGB, claiming that the DHS had withheld information and wasn't being open with the public.[18] When asked about the backlash to the Board, the DGB's (former) executive director, Nina Jankowicz, stated in an interview, "They [DHS] didn't anticipate this fierce backlash and weren't able to mount a transparent, open, and rapid response."[19]

All of this highlights the need to *appear* open, honest, and truthful while engaging in efforts to conceal what is actually happening behind the scenes.

Had Hawley not been offered internal documents, for instance, success would have been more likely. Had the rollout of the DGB been more competently executed, people more assured of privacy, and the functions of the Board and its more controversial aspects obscured, perhaps the DGB would still be operational.

Second, the fate of the Board highlights the need for a firm commitment to the principle of "noble deception." Stated plainly, there are things that government needs to do—for the benefit of the public—that average citizens simply cannot or should not know, both for their own good and for the good of the country. Some things necessary for the maintenance and spread of freedom are too precious to put in front of a critical populace, lest they be destroyed by waves of ignorant criticism.

Journalist Irving Kristol succinctly captured the motivation behind this principle when he noted that "there are different kinds of truth for different people. There are truths appropriate for children; truths appropriate for students; truths that are appropriate for educated adults; and truths appropriate for highly educated adults, and the notion that there should be one set of truths available to everyone is a modern democratic fallacy."[20]

When it comes to complex matters of national security, members of the American public are like children. The family dog isn't going to be euthanized, he's "going to live on a farm." Sensitive data isn't being tracked and stored, it's "being treated with the upmost integrity and with a profound respect for individual privacy."

Information on national security must be concealed from the public. Americans must believe that the material they are given is the full truth, in the same way that the four-year-old believes his old dog is chasing sheep on that farm.

Think of the wonderful work the DGB could have done were it not for the complaining of "privacy advocates" and a meddling senator! Misinformation could have been stopped, support generated, and unproductive discourse avoided while furthering the goals of the national security establishment in the national and global interest. Instead of tracking and countering misinformation for the benefit of the American public, the DGB was nipped in the bud. Instead of creating powerful tools to fight the disinformation tactics of our enemies, we've been set back in the name of "liberty" and "freedom." Instead of trusting the bureaucrats and national security experts who know best, some "civil liberty" dissenters effectively destroyed one of the very things designed to keep liberties safe! Former DGB executive director Nina Jankowicz discussed the backlash to the Board this way: "It's [the backlash] made me a lot less optimistic about the American response to disinformation. This needs to be a wake-up call that things aren't getting better. … Our democratic discourse, the way it is so polarized so, again, childish … leaves us vulnerable to attacks."[21] Childish, indeed.

Though this would make it seem as though controlling information flow will be impossible, do not despair! You still have *many* advantages in this arena. The first, but not the least, is the presence of severe information asymmetries. Information asymmetries exist when one party knows more than another. In matters of national security, there are many information barriers between members of the national security elite and the American public—and even members of Congress. From information generation to the security classification system, there are real hurdles to members of Congress and the general public obtaining any real information about what the security state is doing. Use these asymmetries and these barriers to your advantage to achieve your goals for the greater good. You have immense discretion to decide *what* information to release and *when* to release it. Further, when you know that information is going to be released, you can prepare and set the narrative. The following is an example.

When journalist Michael Grabell sought information about the efficacy of and claims of misconduct regarding the Transportation Security Administration (TSA) and the Federal Air Marshal Service (FAMS), the TSA was able to carefully decide when and how to disclose the requested information and took steps to proactively frame the public messaging. "When the TSA finally responded to my seven-year-old request," Grabell wrote, "it included its own analysis of the data. ... The agency only provided data through February 2012, even though in my last email exchange with the office ... I requested the entire database. ... This has become standard practice for many agencies. By delaying [releasing information] for years, the TSA gets to claim the data it releases is old news."[22]

Nothing of consequence came from the release of this data. Let this be a lesson on how to use your control of information to your benefit!

As the DGB illustrated, being overt and open about the operations being undertaken by the government can quickly turn sour. Instead, you must engage in more clandestine and veiled operations. Thankfully, we can look to our past experiences in this area as well. Multiple forms of entertainment—sports, film, and television—have been invaluable in the past twenty years in bringing the American public into line with the will of policymakers, though most ordinary citizens have no idea that this is happening.

Take professional sports as an example. Nearly every sporting event in the United States—from high school to professional leagues—has some patriotic element. Whether it's the singing of the national anthem, a full-field flag display, or a surprise homecoming of deployed soldiers, do not underestimate the power of sports to generate support for your foreign policy objectives and to assist with the widespread adoption of your narrative. The relationship between sports and the military is a long one, but it has solidified over the past twenty years. In fact, the Department of Defense has actively paid major sports franchises—from professional hockey, football, and baseball teams to the National Association for Stock Car Auto Racing (NASCAR)—millions of dollars to participate in patriotic displays with the goal of fostering support for the U.S. military and, by extension, the interventions in which the military is involved.[23]

In the words of John Collins, senior vice president of marketing and entertainment for the National Football League (NFL), "At the NFL we do two things pretty well. ... We bring people together ... [and] we do a pretty good job of wrapping ourselves in the American flag."[24] This partnership is a remarkable one. Communications scholars have noted how the entanglement of sports and foreign policy can lead to the "equating [of] good citizenship with good fanship. If good fans wear their team's colors and root for their favorite plays in good times and bad, and despite any questionable decision making, then the language of sport in politics may also position citizens to acquiesce to the decisions of their elected leaders. ... [Combining sports with politics] may end up limiting, or even eliminating, the open discussions of policy."[25] A simple example will illustrate these effects.

During National Hockey League (NHL) games, it is common during an intermission to offer a "standing salute" to a member of the U.S. Armed Forces. The entire audience is expected to stand and applaud the military member on the ice. Consider the expectations this creates. A person who, for whatever reason, does not wish to stand will undoubtedly feel socially compelled to do so. Whatever we're paying sports leagues for such a service, it pales in comparison to the benefits it generates in terms of the cultivation of patriotism and collective compliance with the goals of the national security elite.

Don't feel that you are limited to the "big" sports franchises such as the NFL or Major League Baseball (MLB). Despite poor viewership in the United States, similar arrangements have been made with Major League Soccer (MLS) and even the Professional Bull Riders (PBR). In summer 2020, for instance, members of the PBR joined the U.S. Customs and Border Patrol along the U.S. southern border with Mexico and "sprung into action when surveillance cameras spotted undocumented immigrants," in what resembled a dramatic scene straight out of an old Western.[26] Border Patrol has spent more than $5 million on its partnership with PBR to generate support for its mission and to bolster recruiting.[27]

Sports aren't your only option for generating broad support for your policies. Film and television provide opportunities as well. The relationship between Hollywood and the DOD goes back more than a century, when D. W. Griffith's *Birth of a Nation* was completed in 1915 with assistance from engineers and artillery from the United States Military Academy at West Point.[28] The CPI and the OWI, both discussed earlier, had their own influence in the way that war, the military, and the activities of the U.S. government were portrayed on the silver screen.

During World War II, for example, the OWI created the *Government Information Manual for the Motion Picture Industry* (GIMMPI) to "assist the motion picture industry in its endeavor to inform the American people, via the screen, of the many problems attendant on the war program."[29] The GIMMPI left no stone unturned with regard to how filmmakers should portray the war effort. The manual emphasized that studios should use their films to encourage citizens to cooperate with authorities. This included encouraging citizens to volunteer their time to help with the war effort and to buy war bonds. The manual also indicated that rationing should be portrayed as patriotic and that entering black markets to buy rationed goods should be considered aiding the enemy. Viewers of films should be made to see, according to the GIMMPI, that *they* were responsible for the success of the war effort and that *they* had to do their part to ensure that success.

Though we no longer have the OWI or the GIMMPI, this is not to say you lack the ability to profoundly influence what studios put into their films or television shows. After 9/11, for example, officials from the George W. Bush administration met directly with the head of the Motion Picture Association of America (MPAA) and other industry leaders to discuss the ways that the war in Iraq and the broader war on terror would be portrayed. Bush's senior advisor and deputy chief of staff, Karl Rove, presented seven points to the likes of CBS, HBO, and MGM. Each of these points—from calling Americans to national service to the idea that the war on terror was a war of "good against evil"—reinforced the narrative of the Bush White House. This narrative emphasized that the actions of the U.S. government were good, gallant, and in service of the public interest. Supporting the public interest required supporting these policies.[30]

Keep in mind that you have more than the power of suggestion to use as leverage over the film and television industries. You *can* affect Hollywood's bottom line.

For decades, film studios have enjoyed the benefit of using discounted or free military hardware, locations, and personnel for their films in exchange for the DOD having a say in the editorial process. Problematic characters, unflattering plot points, and objectionable lines can be altered or completely eliminated at the request of the

DOD. What if studios don't want to make these changes? They don't get to use the military's equipment. It's that simple.

If film studios want to maximize their profits from films that require the use of tanks, fighter planes, or an aircraft carrier, good luck doing so without the help of Uncle Sam.[31] Over the past twenty years, films from *Elizabethtown*, a "romantic tragicomedy" starring Orlando Bloom and Kirsten Dunst, to the *Transformers* franchise have received support from the DOD.[32] And if Hollywood producers are using government resources for their film project, you can bet they will portray the government positively to their audiences.

This process is not exclusive to full-length films. The same dynamic also works for television shows. Shows such as *Say Yes to the Dress*, *American Chopper*, and *Extreme Makeover* have all benefited from DOD support, along with a cadre of cooking shows, from *Emeril Live* to *Hell's Kitchen*.[33]

This demonstrates that a diverse range of shows can be used to reinforce the narrative that *you* create. *Nothing* is off-limits when it comes to matters of national policy and security.

Do not underestimate the power of this tool.

Without having the slightest idea that it is happening, ordinary people enjoying television after work or on a weekend can be consistently exposed to your messaging. Observe with delight as Gordon Ramsay reinforces over and over and over the need to treat the heroes of our armed forces with the utmost respect and reverence as they work to carry out the policies that keep us safe. If a cooking show contestant can deliver a perfectly cooked filet mignon to members of the National Guard under pressure from a Michelin Star chef, the least watchers can do, from the comfort of their couches, is to support the troops—and their mission.

The narrative must be controlled. Should you fail in this regard, you have no hope for success. Thankfully, between the information asymmetries inherent in democratic politics, the historical knowledge and experience gained from organizations such as the CPI and

OWI, and the long-standing relationships between the DOD and professional sports and Hollywood, you have remarkable advantages in this area.

These alone may not be sufficient, however. There is also the problem of the media.

Luckily, we have a solution for that problem, too.

3

Capture the Media

AS FORMER WHITE HOUSE press secretary Scott McClellan stated, the goal of the White House communications team under the Bush administration was to "win every news cycle."[1] This is the goal now as well.

Setting the narrative and controlling the flow of information about U.S. foreign policy is critical. But *who* will deliver the narrative? We just discussed the importance of sports, movies, and television. The news media is so important, however, that it deserves a dedicated discussion. The takeaway is simple but crucial: you must know how Americans consume information and control—or at least have strong influence over—those communication channels.

In a free society, independent media is necessary. But a free society is not guaranteed without a proactive national security state, so logically the security elites must proactively influence and control media for the greater good.

For the public to be convinced to support U.S. foreign policy objectives, the media must be an integral part of your strategy. Your carefully crafted message will have little effect if it isn't conveyed properly. The media is a respected, long-standing institution in the United States. Many people pay close attention to what the media covers. For instance, in 2022, the Reuters Institute for the Study of Journalism found that about half of all Americans are "very" or "extremely interested" in the news.[2] Nearly a third of Americans access radio, print, or televised news weekly.[3] While overall trust in media has declined in recent years, more than half of Americans still report that they trust their local news station, and a plurality of Americans either trust or have "no opinion" regarding the trustworthiness of all major national news outlets.[4] Harness this trust and use it to achieve your goals.

An overwhelming number of Americans still get their news from traditional media formats—radio and television. In 2020, nearly seven in ten adults stated that they obtained their news "often" or "sometimes" from television. Half said the same of radio. Though newspapers across the country are struggling to stay afloat, a third of Americans continue to get news from traditional papers, at least some of the time.[5]

There are stark differences between age cohorts. A quarter of adults over 65 said they "often" get their news from print publications, while only 3 percent of adults ages 18 to 29 reported doing the same.[6] Around 70 percent of adults ages 18 to 49 said they often turned to smartphones or computers to access news, while only 48 percent of adults over 65 said they often used these devices to access news.[7] Use this information to your full advantage!

You will by no means be the first to use the media to generate support for your policies. In fact, you have many examples to follow. In the previous chapter, we discussed the operations of the Office of War Information and Committee on Public Information in World War I and the collaboration between the government and journalists through the Writer's War Board during World War II. During the Cold War period, we can look to the example set by luminaries such as former secretary of state Henry Kissinger. Kissinger not only personally approved thousands of secret bombing raids in Cambodia in 1969 and 1970 but also approved "the methods for keeping [the bombings] out of newspapers."[8] And there are more recent examples you can look to for relevant strategies and inspiration.

There is perhaps no better illustration of how to leverage the media to generate support for your foreign policies than that provided by the George W. Bush administration. The administration's media tactics in the lead-up to and following the invasion of Iraq should be considered a master class for anyone working in the foreign policy arena. Following the attacks on September 11, 2001, the administration began an intensive campaign to garner support for the invasion of Iraq and the overthrow of the country's leader, Saddam Hussein. The administration used its information advantages, carefully and deliberately crafting messages to generate broad support for the administration's policies. These narratives included claims about (1) Iraq's supposed connection to terrorist groups (namely Al-Qaeda); (2) Iraq's weapons program; and (3) broad, global support for an invasion.

While officials knew at the time that these claims were either dubious or, in some cases, patently false, the desired narratives were conveyed to the public—and the media was integral in conveying those messages. It is worth briefly reviewing each component of the Bush administration's messaging strategy, as a template for future use.

The Bush White House ensured that top officials were available for public comment. This is essential to using the media effectively; availability for comment not only makes it appear that the administration is open and forthcoming with the public but also lends credibility and authority to the message being conveyed. While the Bush administration was selling the Iraq War to the American public, for example, officials such as Vice President Dick Cheney and National Security Advisor Condoleezza Rice were readily available to make media appearances. The House of Representatives Committee on Government Reform found that President Bush, Vice President Cheney, Secretary of Defense Donald Rumsfeld, Secretary of State Colin Powell, and National Security Advisor Rice made a combined fifty-two appearances just to discuss the supposed Iraq–Al-Qaeda nexus.[9]

Do not worry about the veracity of the information your officials convey (though you should always make sure they can maintain plausible deniability if worse comes to worst). Repetition and reputation will typically make the message true in the minds of the public. Vice President Cheney, for example, stated in multiple interviews with *Meet the Press* that Mohamed Atta (one of the 9/11 masterminds) and Iraqi leaders had met in Prague just several months before the attacks—even though the Federal Bureau of Investigation (FBI) had Atta in custody in Florida at the time of the supposed meeting.[10] The fact that the administration knew this information at the time was not of consequence. The message was presented repeatedly and consistently in service to the greater goals of the administration.

The repeated appearances by officials and the repetition of their statements about the Iraq–Al-Qaeda nexus by members of the media had the desired outcome. Two weeks after the 9/11 attacks, CBS and the *New York Times* released a poll of Americans' beliefs about who was responsible for the attacks. A mere 2 percent stated that Saddam Hussein was solely responsible for the attacks. Another 6 percent said that Hussein and Osama bin Laden were together responsible.[11] Compare these numbers to those less than five months after the invasion—after the large media push by the White House. Polling by the *Washington Post* in August 2003 found that 69 percent of Americans thought Saddam Hussein was *personally* involved in the 9/11 terror attacks. Some 82 percent agreed that Hussein had likely aided terrorist groups such as Al-Qaeda.[12] Just as we told you, the Bush administration's effort to link Iraq and Al-Qaeda can serve as a master class in changing public opinion!

Similar results can be found when we examine the Bush team's media narrative surrounding weapons of mass destruction (WMDs). Again, officials were made publicly available; news conferences, interviews, and press briefings were widely televised, quoted, or otherwise disseminated. The message was clear: Iraq possessed WMDs, and these presented a direct threat to the United States, its allies, and global stability and freedom. Consider just a few of the statements offered by former president Bush prior to the U.S. invasion of Iraq.

President Bush began preparing the American public for his handling of Iraq before he was even in the White House. Appearing in a nationally televised GOP debate in 1999, Mr. Bush stated, "If I found in any way, shape, or form that [Saddam Hussein] was developing weapons of mass destruction, I'd take them out."[13] In an interview on *PBS NewsHour* with Jim Lehrer in February 2000, then-candidate Bush rebuked the Clinton administration's handling of Iraq. "I think we need to make it clear ... we expect the [weapons] inspectors back in to make sure [Saddam Hussein is] not developing weapons of mass destruction. ... If we catch him developing weapons of mass destruction in any way, shape, or form, I'll deal with that in a way he won't like." Lehrer clarified, "Like what, bomb him?" Bush replied, "It could be one option. He just needs to know that he'll be dealt with in a firm way."[14]

Just prior to the invasion, in February 2003, President Bush again stated in his national radio address that Hussein had weapons (and linked him once again to terrorist networks): "The regime has never accounted for a vast arsenal of deadly biological and chemical weapons. ... One of the greatest dangers we face is that weapons of mass destruction might be passed to terrorists who would not hesitate to use those weapons. Saddam Hussein has long-standing, direct, and continuing ties to terrorist networks."[15] Other officials, such as Vice President Cheney and Secretary of State Powell, made similar declarative statements in the lead-up to the war.

It didn't matter that the reality was not nearly so definitive. Neither the Bureau of Intelligence and Research nor the International Atomic Energy Agency had found any indication of nuclear or other heavy weapons in Iraq.[16] When it came to the infamous "aluminum tubes" and claims that Iraq had purchased uranium from Africa for the purposes of developing nuclear or other weapons, government officials knew at the time either that such transactions had never occurred or that the materials had been acquired for alternative, legitimate purposes.[17]

But the reality didn't matter for them, and it shouldn't for you. Removing Saddam Hussein from power was the correct action, and the Bush administration used the media to help get the public on board with the policy. Indeed, members of the public largely accepted the narrative they had received (for years at that point) through various media channels. In September 2003, a *Washington Post* poll found that 84 percent of those surveyed thought the Iraqi government was looking to develop WMDs. Some 78 percent thought Hussein's regime already possessed them.[18] More than a decade after the invasion, the narrative pushed by the Bush White House was still the narrative accepted by a large percentage of Americans. Polling conducted in January 2015 found that more than half of Republicans (51 percent) agreed that it was "definitely true" or "probably true" that U.S. forces had found weapons of mass destruction in Iraq. But it wasn't just Republicans. Some 46 percent of Independents and 32 percent of Democrats surveyed answered similarly.[19]

The American public also believed the widely publicized claim that the U.S. government had wide-scale international support to remove Saddam Hussein. A *USA Today* poll conducted just days before the war started in March 2003 found that Americans supported an invasion by a two-to-one ratio—but only if the United Nations granted its approval.[20] A series of other polls produced similar results. Americans would support the war, but only if they felt the United States had foreign support.[21] But neither the fact that the United Nations *did not* support the invasion nor the lack of support from the international community actually made any difference. The reason is that the Bush administration, with the aid of the media, helped to tell a different story. Just as the administration had done with the supposed Saddam Hussein–Al-Qaeda linkage and the purported WMDs, the Bush White House emphasized that the world supported the invasion.

In a widely publicized meeting with British prime minister Tony

Blair in 2002, Bush once again stated that there was widespread support for intervention in Iraq, tying this support to other components of the White House narrative: "A lot of people understand he holds weapons of mass destruction. ... A lot of people understand he is unstable. So we've got a lot of support."[22] "The bottom line," stated former White House press secretary Ari Fleischer during a daily press briefing on March 18, 2003, "a coalition of the willing will disarm Saddam Hussein's Iraq."[23] The "coalition," stated Secretary Powell, was some thirty nations strong, with another fifteen countries that were willing to offer support but "[did] not yet wish to be named publicly."[24]

Only the United States, Britain, Australia, and Spain would supply any significant number of troops. But that didn't matter. Most countries were vocally against the war—including major U.S. allies, such as Germany and France.[25] But that didn't matter. The fact that most of the countries listed in the "coalition" would supply nothing at all made no difference. After all, what could countries such as Rwanda, which in 2003 had a per capita gross domestic product of $704, or Palau, which has a population of less than twenty thousand people, possibly offer to the war effort?[26]

Nothing, of course. But again, that didn't matter.

What *did* matter was that the media broadly put forward the narrative of the Bush White House, over and over again, and the public bought it.

Using the media to generate support for your policies at their outset is necessary, but you will need to employ different tactics if you wish to maintain public support through time. After the invasion of Iraq, for example, support at home eventually waned, and the Bush administration's public relations machine was forced to alter its media tactics. The strategies employed here are again illustrative and ones you should consider for use in the future.

Making members of the administration available for interviews is important. As we just highlighted, having officials willing and able to discuss policy is an effective tool. However, some individuals will not trust "Washington insiders" when it comes to information about U.S. foreign policy.

In that case, you need to find ways to offer "objective" analyses from messengers deemed credible by the public. The Bush White House accomplished this in two primary ways. First, the administration consistently offered various pieces of information—new "leads"—to media outlets. These outlets, anxious to be the first to report on this new information, would quickly turn around and report it, often without determining its authenticity. Later, while giving interviews, officials would cite *the outlet* as the source of the information, even though it actually originated with the White House. This was a great asset to officials, who could claim that the information being discussed—about WMDs, Hussein's supposed terrorist connections, or whatever it was—came from external, objective news sources.[27]

The other way in which the Bush administration provided "objective" information to the public was by offering media outlets access to "experts" who were ready and able to appear on their programs. Throughout the war in Iraq, "military analysts" appeared frequently on programs. These supposedly objective experts, however, would offer White House talking points.

Drawn from the legions of defense contractors and lobbyists, these experts had direct financial incentives to continue the conflict, and so they had an incentive to toe the party line.[28]

This situation has not changed. As we will discuss in more detail later, there are numerous companies and individuals whose financial interests are intimately tied with whatever foreign policy program you undertake. Leverage their interests to suit your own. If they have doubts about the narrative they're supposed to be putting forward, this is of little consequence. "Experts" are easy to come by, and, put simply, they need you to stay in business.

"Some analysts acknowledge they suppressed doubts," wrote journalist David Barstow in the *New York Times*, "because they feared jeopardizing access."[29] You would do well to remember this dynamic and use it to your advantage. The career success of journalists depends on the access *you* provide them, and for many members of the media, these career incentives will trump their commitment to their supposed "principles."

But what is to be done about the media in conflict zones? We recommend the process of *embedding*, in which carefully vetted members of the media are placed with carefully selected deployed troops. During the war in Iraq, for instance, the administration placed with military units some six hundred journalists from a cadre of different media outlets (remember, when you're seeking to saturate the media market, that readers of *Playboy* are typically not also readers of *Reader's Digest*).[30] Embedding journalists has four main benefits.

First, by placing reporters in the field, you effectively limit their ability to assess the overall picture, creating what's been called the "soda straw" effect (the ability to see only a very limited portion of a larger and more complex set of issues). "This effect can be mitigated," write researchers, "to the extent that the public has access to the views provided by many soda straws."[31] You cannot allow that to happen, of course. The journalists will see what you want them to see, and, in turn, their audience will see what you want them to see.

Second, you can keep journalists away from the front lines of the conflict in the name of protecting their safety. This further limits their ability to gather complete information and will make them reliant on the information they receive from you through government-affiliated liaisons to flesh out their stories.

Third, the process of embedding tends to influence journalists' objectivity. Think about it this way: Would you want to report negatively on the people directly responsible for your safety? Of course not.

Fourth and finally, embedding is another way to limit dissent among the press. Journalists, whose careers depend on their relationships with sources and their ability to get the inside story, face strong incentives not to burn bridges with the government.

Does a reporter consistently show U.S. foreign policy activities in a positive light? Perhaps he should be rewarded with new information. But what if some budding Walter Cronkite decides his "thoughtful" or "critical" perspective is more appealing or important? Put him on the next flight home. He can be sure never to receive any privileged information ever again.

Remember: not only do you control the narrative, but you are also the gatekeeper of the information.

Discussing the significant benefits of embedding from the perspective of the U.S. government, journalist Jack Shafer wrote, "The Pentagon officer who conceived and advanced the embedded journalist program should step forward and demand a fourth star for his epaulets. By prepping reporters in boot camps and then throwing them in harm's way with the invading force, the U.S. military has generated a bounty of positive coverage of the Iraq invasion, one that decades of spinning, bobbing, and weaving at rear-echelon briefings could never achieve."[32]

Without a doubt, times have changed since the war in Iraq. The tech giant Facebook, for instance, was not founded until 2004. Twitter (now X) wasn't launched until 2006, and Instagram was only available starting in 2010. Given that they host *billions* of users and that many of these users get at least some of their news from social media, what do you do with these platforms?

This is a tricky but important question.

The Biden administration may provide a useful starting point. During the COVID-19 pandemic, White House officials met with members of these social media platforms to coordinate on "misinformation" and supporting the official narrative on pandemic policies (for example, masking and vaccines). In July 2021, then–White House press secretary Jen Psaki stated, "We are flagging problematic posts for Facebook that spread disinformation." She added, "It's important to take faster action against harmful posts … and Facebook needs to move to more quickly remove harmful violative posts."[33] Members of the government, of course, get to determine what is accurate and inaccurate information, a power that should be extended to all matters of national security.

In the case of COVID, it appears that Facebook was more than happy to comply with these requests. In one redacted email between an unidentified official from the U.S. Department of Health and Human Services (HHS) and Facebook personnel, the coordination between Facebook and government health officials is stated clearly.

It is also clear who was in the driver's seat.

"I know our teams met today," wrote the representative from the tech giant, "to better understand the scope of what the White House expects from us on misinformation going forward."[34] Similar sentiments were reiterated in another email from Facebook to HHS officials.

> We think there's considerably more we can do in partnership with you and your teams to drive behavior. We're also committed to addressing the defensive work around misinformation that you've called on us to address. … We believe our work is paying real dividends. … We're eager to find more ways to partner with you.[35]

Other emails between officials and the tech company suggest that Facebook was enthusiastic when it came to assisting officials in pushing their desired narrative while suppressing information to the contrary. In one email, a Facebook representative suggested to an official at the Centers for Disease Control and Prevention (CDC) that in addition to their weekly meetings, the CDC and Facebook could benefit from "doing a monthly disinfo [disinformation]/debunking meeting." The email suggested that "topics [be] communicated a few days prior so that you can bring in the matching experts." The official from the CDC replied, "We would love to do that."[36]

What a wonderful example of how private social media can be a force multiplier in service to the greater good! And with fast-paced changes in technology, this is only the start.

When it comes to X, the company has long maintained that it works to detect and shut down covert, government-backed accounts. In reality, however, X appears to have been a willing participant in helping officials spread their foreign policy messages. Journalist Lee Fang uncovered a variety of emails and other information about how the company assisted government officials:

> Behind the scenes … the social networking giant provided direct approval and internal protection to the U.S. military's network of social media accounts and online personas, whitelisting a batch of accounts at the request of the government. The Pentagon has used this network, which includes U.S. government-generated news portals and memes, in an effort to shape opinion in Yemen, Syria, Iraq, Kuwait, and beyond. The accounts in question started out openly affiliated with the U.S. government. But then the Pentagon appeared to shift tactics and began concealing its affiliation with some of the accounts—a move toward the type of intentional platform manipulation that Twitter has publicly opposed.[37]

Fang found that X had assisted the Pentagon at least since 2017, when a United States Central Command (CENTCOM) official emailed the company with the request to verify certain accounts that, according to the official, were used to "amplify certain messages."[38] These accounts were in turn "whitelisted." This "essentially gave the accounts the privileges of Twitter verification without a visible blue check [a visual blue mark on accounts for which the company has verified the user]. Twitter verification would have bestowed a number of advantages, such as invulnerability to algorithmic bots that flag accounts for spam or abuse."[39]

Remember that government agencies possess the ability to regulate private entities. And given that Facebook and other tech companies have come under threat of intensive regulation in recent years, it is in their best interests to cooperate with government officials.

Be sure to leverage government's regulatory powers, and be sure to remind the leaders of technology firms that if they fail to cooperate, they might just find themselves subject to more restrictions that could harm their operations and profitability.

As noted earlier in this guide, it is imperative that you make sure that you are publicly viewed as committed to honesty and as the ultimate protector of accurate information. Speaking to a group of press members in the United Kingdom in October 2023, Vice President Kamala Harris stated, "I am clear that one of the greatest threats to democracy is mis- and disinformation. Now, the face of mis- and disinformation is not new. But ... with the evolution of technology ... disinformation can spread quickly."[40] She later continued, "I come at this issue with the experience of having seen how it plays out in real time in our own country and understanding, in particular, how technology can be a tool in the hands of bad actors who intend to upend democratic institutions and the people's confidence in democracies."[41] When asked whether there was anything to be done to combat misinformation, the vice president stated succinctly, "We're going to do everything we can."

Indeed, the Biden administration has worked diligently to cultivate a reputation not just for seeking the truth but also for being purveyors of truth. Consider, for instance, the following exchange between a journalist and the National Security Council's coordinator for strategic communications, John Kirby, at a press briefing on the war between Israel and Hamas:

> Reporter: This administration has fought very hard against misinformation. *You hold the truth as a gospel*, so facts matter, I guess. ... I just want to make sure that you are 100 percent sure of these stories [of atrocities committed by Hamas in their initial attack on Israeli civilians], because they prepare us for what [is] to come. And I have in mind the Iraq War, where notable outlet[s], including many of our colleagues here, disseminated misinformation and governments' lie [*sic*], actually, about the war. There is no comparison here. But I'm saying—
>
> Kirby: There's no comparison—
>
> Reporter: I don't take—
>
> Kirby: Whatsoever.[42]

When the journalist asked how to separate "facts from misinformation," Mr. Kirby quipped that the journalist should register for a course Kirby teaches at Georgetown University and went on to say, "We [the administration] take it very, very seriously—the—the need to be as factual and certainly, truthful as we can possibly be."[43]

Precisely. Be as truthful as you can *possibly* be, as determined by *you* in light of the goals *you* seek to achieve. You are the expert, and you know better than others. This expertise bestows a duty on you to determine what the media and public must know and then take the necessary steps to make sure the media falls in line so that you can achieve your goals of combating illiberalism at home and abroad.

While some of the above may seem distasteful or counter to the ideals of a free society, you must never forget the bigger picture. You are working to maintain and spread liberty and freedom at home and abroad. This requires you to curtail freedom of speech and freedom of the press in order to protect these freedoms, for otherwise they will be lost.

Imagine a world dominated by a secretive, authoritarian government that controls the flow of information and what people can say and think. It sounds horrible, because it is. You must not, and cannot, let this come to pass under any circumstances.

4

Prepare the Sacrifices

YOU ARE ALL familiar with the maxim "You've got to break some eggs to make an omelet."

Sacrifice is often needed in order to achieve your goals, and war is no different. If you want to achieve a global liberal order, sacrifices will have to be made. Absent voluntary cooperation, force will be required.

You must be prepared to use violence, up to and including killing other human beings. Do not be deterred by the use of these illiberal ends to achieve liberal means. Remember, all of this is for the greater good—the best interests of the nation and the world.

In the case of U.S. foreign intervention, the "eggs" you will crack will come from two different "cartons": foreigners and the U.S. military. We'll discuss each of these groups in turn.

In war, enemy casualties are properly expected. As a result of the necessary force, however, collateral damage—the maiming and murder of innocent civilians—will be an unfortunate inevitability. [1]

Luckily for you, most of the innocent people killed in operations abroad will be foreigners. While there is evidence that the American public is at least somewhat averse to civilian casualties abroad, you have several tools at your disposal to help you mitigate these responses or head them off entirely before they become troublesome. [2]

The first of these advantages consists of the information asymmetries and monopoly control over information discussed earlier.

Decisions about when and how to release information about civilian deaths abroad should be aimed at minimizing negative reactions. Remember that the news cycle moves quickly and that members of the public tend to lose interest at an amazing speed.

This is even more true if the event being reported happened sometime in the past. In such cases, the event is old news even as it's being reported. People today see that a dozen members of a wedding party were accidently killed a month ago in Yemen?[3] Or that thirty Afghan pine nut farmers were the unintended victims of a drone strike three weeks ago?[4] "Oh, that's a shame," the viewer *might* think, as she returns to making breakfast for her children. She probably won't think about it twice, even if she thinks about it once.

The second advantage you have is the power of strategic concealment. If you are concerned about the type or number of casualties, make the casualties *disappear*. One way of doing so is through simple wordplay. How you *define* a "combatant" or a "civilian" isn't as clearcut as you might assume.

Consider, for example, the term *military-age male* (MAM). This term, though not formally defined by the military, is used frequently in counterinsurgency efforts to justify a killing—and to exclude such killings from tallies of civilian casualties.

Does a male appear to be between the ages of fifteen and seventy-five? Does he look "threatening"? Does he move after being shot at from a helicopter? Good—that means he was a combatant! His removal was a necessary part of your operation, and his demise conveniently moves from the "unacceptable death" category to the "acceptable death" category.

Discussing his time in Vietnam, former secretary of state Colin Powell captured this idea clearly:

> If a helo [helicopter] spotted a peasant in black pajamas who looked remotely suspicious, a possible
> MAM, the pilot would circle and fire in front of

him. If he moved, his movement was judged evidence of hostile intent, and the next burst was not in front, but at him. Brutal? Maybe so. But [someone I knew] was killed by enemy sniper fire while observing MAMs from a helicopter.[5]

Defining away civilian casualties in the "military-age male" category was a highly effective method under the Bush administration in Iraq and Afghanistan and in the broader war on terror. When it came to the use of drone strikes under the Obama administration, this reframing of casualties versus combatants was invaluable. Not only were any MAMs in a strike zone fair game and excluded from civilian casualty counts, but so were those individuals *around* those being targeted. As one official stated, "Al Qaeda is an insular, paranoid organization—innocent neighbors don't hitchhike rides in the back of trucks headed for the border with guns."[6]

"The Military-Aged Male category is not synonymous with 'combatant,'" writes political scientist Sarah Shoker, "but marks boys and men for differentiated treatment in conflict zones. … Male bodies are used as a shorthand for 'combatant' when assessing the collateral damage count."[7] She goes on to note how, even when it was revealed that some civilian casualties may have failed to be counted due to the MAM category, the media quickly moved on: "This revelation in *The New York Times* attracted much short-lived attention."[8]

See how quickly your civilian casualties can disappear when you engage in a bit of creative accounting?

Unfortunately, you won't be able to define away all the collateral damage. While foreigners who happen to be male can be easily dispatched and explained away, women and children must be handled more delicately. Well, at least their *deaths* must be handled delicately.

Thankfully, we once again have history to draw upon. Consider the case of My Lai in 1968, in which U.S. military personnel gang-raped and murdered some five hundred civilians—including women and children—before burning the village to the ground. An unfortunate and unacceptable event to be sure, but then regrettable things do happen in war.

In such extreme cases, complete concealment is the best option.

Using the aforementioned information asymmetries and the media strategies discussed in the previous chapter, you should make every attempt to keep the information from ever seeing the light of day. The war crimes at My Lai were kept secret for nearly two full years. If nothing else, such a lag will give you time to prepare for the moment when any information is released.

When that happens, a tried-and-true explanation is the classic "a few bad apples" argument. This approach blames the unfortunate event on some rogue rank-and-file soldiers or some ill-equipped or otherwise deficient officer(s) who failed to uphold the standards of the U.S. military. It is crucial to publicly emphasize that these situations are not the norm and that the actions of the guilty in no way reflect the actions of the broader organization or the goals of the intervention itself. The goal is to sanitize war by minimizing the brutality and dehumanization seen by members of the public, lest they lose their appetite for foreign adventures abroad.

This was precisely the method employed when the My Lai story was made public.[9] Similar statements were made following the release of photos showing U.S. military personnel torturing and humiliating detainees at Abu Ghraib prison in Iraq in 2004. Officials repeatedly referred to the military members taking part in the torture as "a few bad apples." Speaking about the incident, President Bush stated, "Abu Ghraib ... became a symbol of disgraceful conduct *by a few American troops who dishonored our country and disregarded our values.*"[10]

To see how well collateral damage can be concealed, consider that we *still* don't know, and will likely never know, the true number of casualties in the war on terror.

No one is sending people into these conflict zones to accurately account for the dead and wounded. And no one with any sense is going to believe the propaganda coming out of places such as Yemen and Pakistan, where domestic governments and groups will undoubtedly inflate the numbers of civilians killed for the sole purpose of making the United States look like a monstrous bully. It is imperative for you to discredit foreign sources by making clear that they are anti-American and that they dislike freedom and your global project of spreading liberal democracy to the world.

Still better for you is that even when credible bodies report on civilian deaths, these stories tend to receive little attention from the public or from members of Congress. One report by the United Nations, for example, found that allied forces in Afghanistan had killed more civilians than members of the Taliban, but the U.S. media largely ignored this information—and the public did, too. When the United States relaxed its rules of engagement for air strikes in Afghanistan in 2017, there was a 330 percent increase in civilian casualties—but only those truly interested in the subject are aware of this change in policy and its results.[11]

Other injuries that occur as the result of conflict are even less likely to be reported. For example, after the United States used "thermobaric" weapons in the heavily populated city of Fallujah, Iraq, in 2004, about half of all children born in the city arrived with some serious birth defect (compared to 2 percent before the U.S. government's military operations). While 10 percent of pregnancies before the siege ended in miscarriage, one in six women afterward lost their unborn children. These outcomes, likely caused by exposure to materials such as lead, mercury, and depleted uranium in the weapons used by the U.S. military, were scarcely reported and remain unknown to most Americans.[12]

This indifference among large segments of the American public is a wonderful tool that you should leverage to the greatest extent possible.

Outside of foreigners, the other group who will sustain casualties will be members of the U.S. military. Here you must tread carefully, as U.S. citizens are much more invested in the lives of their service members than in those of some "foreigners" thousands of miles away. While members of the public will tolerate some military casualties if they believe the war is worthy and winnable (this is one reason you *must* always emphasize that your goals are noble and that America is winning!), there is clear evidence that too many military deaths can negatively impact public support for military interventions.[13]

Therefore, you need a plan for dealing with inevitable military losses.

The best method for dealing with military casualties is to use the deaths to generate support and patriotic fervor. There is perhaps no better example of this than the case of former NFL player Patrick "Pat" Tillman. Tillman left his lucrative football career with the Arizona Cardinals and enlisted in the military following the 9/11 attacks, joining the elite U.S. Army Rangers. In April 2004, Tillman was killed in Afghanistan. It was immediately apparent to key leaders in the army that he had been a victim of fratricide, killed by his fellow Rangers.

Though Tillman had rebuffed attempts to use his story of football-star-turned-patriotic-soldier while he was alive, the Bush administration seized the opportunity to use his death as a means to generate support for the war. Instead of reporting that his death was a result of "friendly fire," the White House and the military reported that he had been killed in a firefight with Afghan insurgents.

By all accounts, Tillman's "heroic" death was a rallying point for the American public. A "Weekend Media Assessment" created several days after it was announced that Tillman had been killed stated that "the Ranger Tillman story had been extremely positive" and that his death had generated the most Army media coverage "since the end of active combat."[14]

Although the truth of Tillman's death is now known (the actual circumstances surrounding his death were made public the Friday before a major holiday, when most reporters had already left for the day; remember, *you* control when information is released), he remains a bipartisan focal point and a tool to generate support for the U.S. military. In September 2017, former president Donald Trump shared a photo of Tillman on X (formerly Twitter), stating, "He fought 4 our country/freedom."[15] He was making a comparison between Tillman and those NFL players choosing to kneel during the national anthem as a form of protest. More recently, in January 2021, former Arizona Cardinals defensive end J. J. Watt paid tribute to Tillman. After visiting Tillman's locker, Watt stated on X that it was "a true honor to walk the same halls." Accompanying the post was a picture of Watt standing by the display case featuring Tillman's jersey and other gear.[16] When someone responded to the picture stating that Tillman wasn't a particularly good player and was killed by his own men, the response was swift. "WOW!! are you seriously talking bad about Pat Tillman? He is a HERO … At one point Tillman turnt [*sic*] a 9 million dollar deal from Rams down out of loyalty to Cards. He then went to War."[17] Another stated, "Pat gave up millions of $$$$ to fight for OUR freedom."[18] Still another chimed in, "Wow you're a next level fucking piece of shit."[19] Nearly twenty years after his death, Tillman's story continues to generate strong patriotic reactions from the American public, demonstrating the power of linking soldier deaths with the rhetoric and symbolism of heroism and patriotism.

But what is to be done about those killed who *aren't* as inspiring as Tillman? What do you do about those men and women who come from small-town America—your "average Joes and Janes"?

In these cases, it is helpful to employ the sunk cost fallacy. A sunk cost refers to an outlay (monetary or nonmonetary) that, once made, cannot be recovered. While sunk costs should not factor into decision-making (you cannot recover sunk costs, and these costs will remain the same regardless of any subsequent decisions), they often do, because many people have a difficult time letting go of the past.

For instance, you often see business owners refuse to shutter their doors because they've "spent so much money" or couples who remain married because they've "been together so long." The business owner cannot recoup his money. The couple cannot recoup their time. Nevertheless, many people allow such past expenditures to influence their decisions to continue doing the same thing. You should use this human tendency to leverage the sunk cost fallacy for your own benefit in matters of military intervention and loss of military lives. It is easy to do so, because the focus is on the lives of American soldiers and because, if you have heeded our earlier advice, you will have invested effort in effectively fostering public support, in the name of patriotism, for the intervention.

Indeed, it is imperative that you convince members of the public that they *should* consider the military lives lost when deciding whether or not to continue offering their support to your foreign policy. If they withdraw their support and if the policy of intervention is reversed, then all the sacrifices will have been in vain. Sons and daughters will have died for nothing—and that cannot happen. When implemented effectively, the sunk cost fallacy can be leveraged to turn death into a justification for more resources and troops to fight abroad, even if that means the loss of more lives in the future.

President Bush employed this strategy effectively. Consider his speech at Fort Bragg in June 2005:

> I thank our military families; the burden of war falls especially hard on you. In this war, we have lost good men and women who left our shores to defend freedom and did not live to make the journey home. I have met with families grieving the loss of loved ones. ... We pray for those families. *And the best way to honor the lives that have been given in this struggle is to complete the mission.*[20]

But it wasn't just former president Bush who used this strategy. When President Obama announced renewed efforts in Afghanistan and Pakistan in 2009, he employed a similar tactic:

> Many people in the United States ... have a simple question: What is our purpose in Afghanistan? After so many years, they ask, why do our men and women still fight and die there? ... So let me be clear: Al Qaeda and its allies—the terrorists who planned and supported the 9/11 attacks—are in Pakistan and Afghanistan.[21]

He later continued, "The American people must understand *that this is a down payment on our future*—because the security of America and Pakistan is shared. ... I've already ordered the deployment of 17,000 troops. ... These soldiers and Marines will take the fight to the Taliban ... and give us greater capacity.[22]

This appears to have had the desired effect. For example, Matt Cavanaugh, an active-duty military strategist, wrote in 2015, "Judgments about the value of an individual sacrifice are sacrosanct, to be protected, and limited only to those closest to the deceased. ... We cannot and will not know the value of a soldier's sacrifice. ... We owe them reverence and care with our words; in essence, their noble deaths have elevated them beyond the pigpen of politics. That is why these dead shall not—shall *never*—die in vain."[23]

Following the withdrawal of U.S. troops from Afghanistan, many writers echoed similar sentiments—that military personnel either did not or *cannot* have died in vain.[24]

The very thought that individuals in the military could have died without achieving anything of consequence is simply unfathomable. Take this as a lesson for how you can effectively leverage past deaths to send more soldiers into harm's way in the name of achieving your goals for the greater good. If used effectively, past deaths will reinforce public support for the intervention, rather than undermine it.

While the costs to military service members and veterans are certainly high—from mental health to life and limb—they can easily be offset or ignored. A recent *New York Times* investigative report revealed that a large number of soldiers sent to combat the Islamic State in 2016 and 2017 returned home with severe mental issues, including depression, hallucinations, and suicidal tendencies.[25] The article notes that the cause of these issues was that the troops were tasked with firing tens of thousands of explosive shells at their targets. A 2019 report by the Marine Corps found that the weapons were causing harm to the troops tasked with using them.[26] The report was largely ignored, both privately and publicly, by the U.S. military and officials. The concerns of some troops suffering perverse effects were dismissed. Others were given psychotropic drugs to treat their symptoms. Still "others who started acting strangely after the deployments were simply dismissed as problems, punished for misconduct and forced out of the military in punitive ways that cut them off from the veterans' health care benefits that they now desperately need."[27] As this shows, ignoring the problem is also an effective and sufficient strategy.

However, should you feel any scruples about the costs to soldiers or wish to assuage any public concern, there are steps you can take. First, you should work to ensure that the members of the armed forces are elevated in their social status.

Soldiers are heroes. Full stop. To think otherwise is a grave offense.

Fortunately, many Americans already hold this view. Consider a 2018 YouGov poll that found *half* of Americans believe that "all those serving in our armed forces should be described as heroes, whatever their role or experience."[28] That's correct. Pull your fellow Marines out of enemy fire? Hero. Conduct reconnaissance in Yemen with a UAV while sitting in an Air Force base in Nevada? Hero. Bring the sergeant major his sandwiches? Hero. To solidify this status, consider strengthening the Stolen Valor Act of 2013, which made it a felony to obtain "money, property, or other tangible benefit" by claiming to have received various military honors.[29] A felony has more gravitas in the eyes of the public. Harsher penalties will further convey the severity of such an offense.

Second, you should remember and highlight that government provides health care and other benefits for veterans. Be sure to invest resources in marketing the Veterans Affairs (VA) hospitals as the pinnacle of the U.S. health care system. Never mind that these hospitals often have emergency room wait times that are longer than those in non-VA hospitals, or that VA patients are less likely than non-VA patients to say that "medical workers treated [me] with dignity," or that half of VA hospitals have higher rates of intestinal or bloodstream infections than non-VA hospitals.[30]

In addition to health care, be sure to tout the military's generous disability benefits. A service member who loses both arms, for example, is entitled to full disability benefits of just over $3,600 a month, or $43,400 a year.[31] If he has a wife and two minor children, this amount increases to a more generous $4,072.12 a month, or just over $48,800 a year.[32] That's a full 72 percent of the median household income of the United States!

If it seems that the U.S. government may be on the hook for large sums to pay for veteran care, don't despair. The VA makes sure that only the most deserving veterans obtain these benefits. In fact, 40 percent of all VA disability claims are denied. Veterans who apply without assistance have a 26.2 percent chance of having their claims accepted.[33] As a result, many disabled veterans seek out attorneys or other experienced groups to help with their claims. This gives them a 40.9 percent chance of being approved.

And if they aren't fully disabled? Rates are adjusted accordingly. People don't *really* need all their fingers to work. Toes are even less important. Losing two, three, or even four toes entitles a veteran to 10 percent disability benefits, unless one of them happens to be the great toe. In that case, it's 20 percent.[34] Regardless, if a veteran disputes a denial, it takes an average of seven years for the matter to be resolved.[35] Thank you for your service; we appreciate your patience.

We would be remiss if we failed to discuss *how* to entice individuals to be willing to join the military. Since compulsory service is no longer a politically viable option (at least for now), how do you convince individuals to be willing to risk life and limb for your cause?

For some, a sense of patriotic duty will compel them to join the armed forces and fight on your behalf. One recent study found that many recruits join out of a sense of patriotism. "Many Americans continue to subscribe to an idealized image of soldiers and officers as self-sacrificing patriots," the study's authors write. "Belief in the service member as, first and foremost, an exemplary patriot and citizen can be found across the U.S. political spectrum."[36] Cultivating a belief in self-sacrifice for the greater good—we must fight collectivism in defense of individual freedom, after all—is one of the most effective tools available to you.

But you don't have to rely on patriotism or a sense of civic duty alone to recruit the instruments who will carry out your foreign policy. Socioeconomic and other factors will help, too. Never underestimate what people will do for a steady paycheck, assistance with future home purchases, or health insurance. Want free college? That too can be offered to recruits in exchange for their willingness to be sent off to kill and be killed at the behest of Uncle Sam. Don't forget to mention such things as cheaper on-base groceries, discounted auto repair, and all those perks civilian retailers offer to members of the military.[37] Post-traumatic stress disorder is a small price to pay for deals and celebrations on Veterans Day.

Here's just one example of how we target and attract these cogs in the military machinery. An analysis of Connecticut high schools found that Army recruiters are present in some form on a weekly basis on some campuses—*some* campuses. Comparing two similar-size high schools, the authors found that the school serving wealthier students (i.e., only 5 percent of the student body was eligible for free or reduced-price lunch) received a mere four visits from recruiters. In contrast, the poorer school (where about half of students were eligible for lunch aid) saw ten times as many visits from recruiters.[38] A 2017 study by the RAND Corporation found that Junior Reserve Officers' Training Corps (JROTC) programs (discussed in more detail below) are "well represented among schools serving economically disadvantaged populations," regardless of how that disadvantage was measured.[39] Always be sure to emphasize that service and self-sacrifice of the individual for the greater good is both patriotic and a way out of poverty due to the benefits we offer—unless, of course, the recruit dies before cashing out.

Another option to bolster the ranks is to recruit noncitizens. About 45,000 noncitizens are currently working in the U.S. armed forces. Many are allured by the prospect of citizenship, though new policies have made this path more difficult.[40] Latinos, while grossly underrepresented as officers, are overrepresented among enlisted personnel.[41] This strategy is actively being pursued due to recruitment shortfalls. In fact, the Army runs commercials in Spanish. "You have to recognize that the mother is a dominant influence in Latino families in terms of big decisions," Major General Dennis D. Cavin said of appealing to "the Latino market."[42] The incentive for immigrants is clear: agree to fight in the name of the United States, and we will consider fast-tracking you for citizenship. As journalist Lolita Baldor put it, "Uncle Sam wants you and *vous* and *tu*."[43]

Women are another group you'll want to target. To be clear, this effort will be different from what was done the last time the national security elite needed to recruit a large number of women, which was during World War II.

Gone are the days when you can mandate "Victory Red" lipstick for your recruits of the fairer sex. Gone are the days when you can have women in the military without paying or offering them the same military benefits as their male counterparts.[44] The fact of the matter is that you need women to accomplish your goals. As U.S. Naval War

College professor Lindsay Cohn put it, "I think it's important to say this out loud: we're going to need more women."

She continued, "The difficulties we face with some of the young male population ... in terms of drug use and ... criminal records—are not as big problems in the female population."[45] Researchers from RAND and elsewhere have curated a number of suggestions for boosting and retaining female recruits, from changing advertising strategies to employing more female recruiters.[46]

This will likely not be an easy task. Women are less inclined to enlist than men and, as it turns out, particularly put off by the prospect of being sexually assaulted. The aforementioned RAND study noted that, when conducting focus groups with new military recruits, the researchers "deliberately did not ask specific questions about sexual harassment and sexual assault ... due to concern from the [military] services that raising the topic ... could have unintended negative effects on the recruits' desire to join the military." The same report highlighted that family members and other "influencers" frequently brought up the prospect of sexual assault when their daughters, girlfriends, and so on brought up the idea of enlisting. Of course, this concern has some relevance. The military's most recent report on the prevalence of "unwanted sexual contact" showed some thirty-six thousand reports of sexual misconduct—mostly against women.[47] The DOD estimates that some 5.5 percent of female Air Force personnel and 8.4 percent of female Army members experienced a sexual assault or unwanted sexual contact. The Navy saw higher estimates of 10.1 percent. The Pentagon estimates that 13.4 percent of female Marines have either been sexually assaulted or experienced unwanted sexual contact.[48]

But not to worry, you can assure female recruits and their parents that, of the eighty-two recommendations made to the DOD by an Independent Review Commission set up to investigate the problem of sexual assault in the military, the military has said it's implemented twenty-one.[49]

When asked how they handle concerns over sexual assault, military recruiters have said they've emphasized the policies and procedures in place to combat sexual assault.[50] What else needs to be said? Regardless of gender, you may also consider starting the recruitment process early. Thousands of high schools throughout the United States offer students JROTC classes. Started by the National Defense Act of 1916, the JROTC's mission is to "motivate young people to be better citizens." Part of their training to be better citizens, of course, rightfully consists of a strong introduction to the U.S. military. In many cases, this includes wearing military uniforms, taking orders, and performing drills. Every year, about 314,000 young people learn about the possibility of becoming a part of the U.S. armed forces.[51]

In some cases, school systems are working on your behalf! JROTC subsidizes instructor salaries—so the school saves money on personnel, and you get to make an early sales pitch to impressionable youth.

In 2022, the *New York Times* found that "thousands of public school students were being funneled into the [JROTC] classes without ever have chosen them … or [were] automatically enrolled."[52] The same report stated that "a review of … more than 200 public records requests showed that dozens of schools have made the program mandatory … including schools in Detroit, Los Angeles, Philadelphia, Oklahoma City and Mobile, Ala. A vast majority of the schools with high enrollment numbers were attended by a large proportion of nonwhite students and those from low-income households."

These programs have their critics, but never doubt their efficacy: they are crucial for replenishing the military ranks while fostering a culture of compliance around your foreign policy goals.

According to James L. Jones, former commandant of the U.S. Marine Corps, "The value of [JROTC] is beyond contest. Fully one-third of our young men and women who join the Junior ROTC program wind up wearing the uniform of a Marine."[53] Admiral Jay L. Johnson, former chief of naval operations for the U.S. Navy, echoes these sentiments: "Even if [the number of JROTC cadets who enlist] is only 30 percent, you know, that's a good number. Think about what we get out of the other 70 percent. They have exposure to us. They have exposure to the military. ... That's a powerful tool."[54] A powerful tool indeed. One of the students interviewed in the *Times* article stated that she and other students regularly heard from recruiters who "pitched the idea of signing up for the military in order to get help paying for college."[55]

While you will need to pay attention to civilian and military causalities, do not let them distract you from the task at hand. To preserve and expand freedom, you *must* be willing to sacrifice others for the greater good.

Those individuals abroad who perish won't be able to enjoy the benefits of a freer society. Depending on how long it takes, their children and their grandchildren may also miss out. But their *great*-grandchildren and beyond may appreciate their sacrifice. Those U.S. service members who are lost or injured will have forfeited their lives or health in the name of spreading freedom. Their families and loved ones will suffer. But never forget that freedom requires that individuals must make sacrifices for the collective good. And, if they're like Pat Tillman, their deaths may prove to be one of the most powerful tools you have to generate ongoing support.

When it comes to thinking about casualties, consider the words of General Tommy Franks. When discussing the number of civilians killed by U.S. forces in Afghanistan, he said, "We [the U.S. military] don't do body counts."[56] You shouldn't either.

Remember, *break the eggs*; the omelet of freedom requires it.

5

Sacrifice Liberty in the Name of Liberty

THEY SAY THAT if you love something, you should set it free. If it's meant to be, it will return. This is how you must think about individual liberty.

To protect and expand freedom at home and abroad, the people of the United States must be willing to sacrifice liberty to their protectors—the national security elite. Once our benevolent hegemony has established a free and prosperous global society, these sacrificed liberties will bountifully return.

This reality must be accepted, voluntarily or by force. Public messaging is crucial: "extreme" measures are temporary necessities to protect individuals' well-being and that of their loved ones. There are no alternatives. If the messaging is done correctly, the public will accept these changes, sometimes with enthusiasm. A wonderful contemporary example of the attitude that must be cultivated relates to the Transportation Security Administration (TSA). Despite searches of property and physical inspections of their bodies, most airline passengers have no qualms about these reductions in liberty.

In discussing enhanced security requirements in a news segment, one passenger cheerfully states, "Happy to comply and happy to be safe." Another says, "It's a pain, but if it keeps us safe, I'm willing to do it."[1] This attitude of acquiescence and proactive support is precisely what must be cultivated.

TSA messaging has been so effective since 9/11 that as of 2022, some 79 percent of Americans said that prioritizing "screening for security over saving travelers time and money" is the right call.[2] This public support exists despite the fact that these reductions to liberty have done little to nothing to actually improve security.[3] Further, most American citizens have no idea that they're just as likely to be killed by a comet as they are to die in a terrorist attack.[4] They similarly don't know that the TSA often fails its own tests—in which officials attempt to sneak weapons and other dangerous materials through security—an astonishing 95 percent of the time.[5]

This demonstrates how effective messaging can maintain support for policies that run counter to your stated goals. Even if your policies are ultimately ineffective, they are important to implement to produce observable indicators that you are "doing something" to protect the public from "global threats." And even in the face of failure, your policies must be maintained to avoid the appearance of fallibility.

This is why constant messaging about ubiquitous and general threats, as well as the importance of your policies for safety, is key.

You might also cultivate the narrative that liberal societies are especially vulnerable to terrorism. You can do so by emphasizing to the public that liberal democracies allow for freedom of speech and association. By allowing freedom of thought, speech, association, and movement, democracies provide radicals the chance to speak, network, recruit, and generate funding for their activities.[6] It is important to tell the public that by allowing these activities to continue unabated, we risk allowing terrorists to kill innocent people while destroying our society. Liberties must be curtailed in the name of freedom and safety.

Undoubtedly, there will be speed bumps to the efficient implementation of your policies—for example, the courts, Congress, watchdog groups, and "civil libertarians." When it comes to reductions in the core freedoms mentioned above, pushback by some is inevitable. It is critically important, therefore, that this dissent be minimized, dismissed, or punished. As we mentioned earlier, you should actively conceal your actions to the largest extent possible. When you cannot, you should undermine the legitimacy of critics and seek to maneuver around legal constraints to achieve your desired ends.

Fortunately, some scholars have curated a simple framing that you can easily sell to the public.[7] And luckily for you, the public has already bought it. Here is how it goes.

There is a simple trade-off between liberty and safety. The government can provide its citizens with some mix of liberty and safety. Expert officials will select the optimal mix of these two items. To have more safety, citizens must accept a reduction in their liberties. More liberty, meanwhile, requires citizens to relinquish their safety and the safety of their families.

For you, this means that when faced with some threat—jihadi terrorists, homegrown extremists, climate change, a pandemic, and so on—the public must be convinced that government is the sole solution to the problem and that reductions in freedom are necessary for the maintenance of safety.

How individual liberties need to be reduced will depend on the specifics of the conflict. Several examples are illustrative. After the attack on Pearl Harbor on December 7, 1941, the United States entered World War II against the Axis powers. Having been attacked on their own soil, the government and the general public were understandably concerned about domestic security threats—particularly on the West Coast. In February 1942, President Roosevelt issued Executive Order 9066. "The successful prosecution of the war," reads the order, "requires every possible protection against sabotage."[8] The order authorized the secretary of war to designate "military areas" from which "any or all persons may be excluded, and with respect to which, the right of any person to enter, remain in, or leave shall be subject to whatever restrictions the Secretary ... may impose in his discretion."[9] The secretary determined it was necessary to remove some 117,000 Japanese Americans, including U.S. citizens, who were sent to internment camps for the duration of the war.[10]

This prohibition on free movement and the reallocation of the homes and businesses of these suspicious agents was undertaken for the greater good. Americans tend to focus on individual rights and liberties. And while that is nice in theory, in practice the individual must often be sacrificed for the collective.

Other past scenarios required different restrictions on liberty. The Cold War, for instance, required careful tracking of dangerous ideas and the extremists who held and discussed these ideas. The House Un-American Activities Committee (HUAC) was created to specifically investigate claims of disloyalty on the part of U.S. citizens. The effects of this committee are clear. For instance, actress Marsha Hunt was effectively kept away from the silver screen after she attended peace rallies and other suspicious meetings. Speaking of her career after being observed by the HUAC, she said, "I'd made 54 movies in my first 16 years in Hollywood. In the last 45 years, I've made eight. That shows what blacklists can do to a career."[11]

Famed actress Lucille Ball was similarly investigated, and for good reason. Ball had registered with the Communist Party in 1936.[12] She also married Desiderio "Desi" Arnaz, a *Cuban*-American actor—a major red flag. Recognizing that the sponsors for their television program (and their careers) could be gone in an instant, Ball and Arnaz were quick to fall in line, voicing their support for the United States and condemning communism. Arnaz stated before filming an episode of *I Love Lucy*, "We both despise communists and everything they stand for." Referring to Lucy's hair, he continued, "That's the only thing red about her, and even that's not legitimate."[13]

While some may decry the HUAC as, ironically, "un-American," the reality is that investigating and questioning people about their affiliations and political activities was necessary to keep Americans safe. Some ideas are just too dangerous.

Those in the spotlight, like Hunt and Ball, have a patriotic duty to uphold American values and voice them publicly as you, their political masters, demand. The key is to create a system of fear among the general public about the potentially severe consequences of deviating from unquestioning support of the government and your polices.

During Vietnam, the Nixon administration arrested some twelve thousand antiwar protesters.[14] Although detaining U.S. citizens for exercising their right to assemble and petition the government may *seem* unconstitutional, consider the following facts.

Convinced that previous tactics had been unsuccessful, a group called the May Day Tribe assembled and organized a blockade of major traffic routes in the nation's capital; their slogan was "If the government won't stop the war, we'll stop the traffic."[15] In such instances, what are officials supposed to do? Preventing government officials from getting to work means preventing the very work that keeps citizens safe. In retrospect, this mass arrest was clearly the only choice, not only to punish the perpetrators but also to discourage future dissent against the war. In more contemporary times, large-scale surveillance operations have proved necessary in the ongoing fight against terrorism, including observation of civilians and world leaders.[16] Let these and similar experiences be lessons to others who wish to dissent against a government committed to upholding and spreading liberal values at home and abroad!

Other post-9/11 examples show other ways in which freedoms may necessarily be curtailed. Consider "material witness" laws.[17] A material witness is a person believed to have information that is critical (material) to a criminal proceeding. Before the start of the war on terror, material witnesses who refused to testify could be arrested, "in narrow circumstances, to secure the testimony of a witness who might otherwise flee to avoid testifying."[18] This law, enacted in 1984, included restrictions on jailing. Material witnesses could be held only for as long as it took to depose them or have them testify. This changed, however, following the 9/11 terror attacks. The Department of Justice cleverly reinterpreted the law to hold, not only witnesses, but those suspected of engaging in terrorist activities.

The American Civil Liberties Union (ACLU) reported that between September 11, 2001, and 2015, at least seventy men living in the United States—including seventeen U.S. citizens—had been held in the U.S. federal prison system under material witness laws. While the ACLU points out that these men were put into prison without a showing of "probable cause" and that the Department of Justice "refused to respect fundamental constitutional rights of detainees, including the rights to be notified of charges, to have prompt access to an attorney, to view exculpatory evidence, and to know and be able to challenge the basis for arrest and detention," the purpose of this reinterpretation of the law is clear—safety is paramount.[19] A temporary inconvenience in the loss of freedom for fewer than one hundred people is a very small price to pay for the greater national good.

According to the same report, seven of the men detained were ultimately arrested on terrorism charges. A full *10 percent* of those held were charged. Even those who weren't charged weren't deprived of their liberties for too long. Those who were imprisoned typically spent only a month or two in prison. Only one spent more than a year behind bars.[20] Of course, if these people had not engaged in suspicious behaviors, they would not have been inconvenienced at all.

Reductions in individual liberties are so important for the preservation of safety that they were included throughout the USA PATRIOT Act. Section 802, for instance, redefined terrorism to include *domestic* as well as international terrorism. The Act defines terrorism more broadly than before, including activities that "involve acts dangerous to human life" or activities that "appear to be intended (i) to intimidate or coerce a civilian population; (ii) to influence the policy of a government by intimidation or coercion; or (iii) to affect the conduct of a government by mass destruction, assassination, or kidnapping."[21] As the ACLU noted, this definition "is broad enough to encompass the activities of several prominent activist campaigns and organizations. Greenpeace, Operation Rescue, Vieques Island and WTO protesters … have all recently engaged in activities that could subject them to being investigated for engaging in domestic terrorism."[22]

This expanded definition could prove to be *remarkably* helpful in the future, as you will be able to easily investigate those who may harm or hinder your national security policy goals. As an added benefit, those who actively dissent may be deterred by a fear of being arrested on terrorism charges. Those who choose to still engage may be investigated under this expanded definition.

Your expanded powers do not end there. Section 806 of the USA PATRIOT Act changed the civil asset forfeiture laws. Under the new law, the government has the authority to seize the following:

> All assets foreign or domestic—(i) of any individual, entity, or organization engaged in planning or perpetrating any act of domestic or international terrorism … against the United States, or their property, and all assets, foreign or domestic, affording any person a source of influence over any such entity or organization or (ii) acquired or maintained by any person with the intent and for the purpose of supporting, planning, conducting, or concealing an act of domestic or international terrorism … against the United States or their property or (iii) derived from, involved in, or used or intended to be used to commit any act of domestic or international terrorism … against the United States, citizens or residents, or their property.[23]

In essence, "the government can seize … assets on the mere assertion that there is probable cause to believe that the assets were involved in domestic terrorism. The assets are seized before a person is given a hearing."[24] It gets even better! In order to seize the assets permanently, the government does not have to prove beyond a reasonable doubt that the assets were used in promoting or engaging in terrorism; the government needs to show only that a preponderance of the evidence supports that finding.

This will be quite useful in quieting dissension. The ACLU notes that "the U.S. government can bankrupt political organizations it asserts are involved in terrorism."[25] This view is far too narrow, for think of the good you can do by wielding this power for the common good. Moreover, since the due process of criminal cases is not required in these instances, you have much more flexibility to hinder those who would do harm to the United States or interfere with your policies.

Any liberty lost is a minor sacrifice to keep everyone safe.

Another example from the war on terror is the use of "targeted killings" of individuals, including U.S. citizens, by the government outside of formally declared armed conflicts. Critics have pointed out that these killings violate U.S. and international laws against assassinations without due process of law. But these critics once again fail to see the big picture. Consider, for example, the case of Anwar Al-Aulaqi, Samir Khan, and Abdulrahman Al-Aulaqi in 2011. Samir Khan, a Saudi-born U.S. citizen, wrote "the most influential bomb-making manual since *The Anarchist Cookbook*" and was "an uncommonly gifted propagandist" who recruited English-speaking jihadists.[26] Anwar Al-Aulaqi, a "U.S.-born radical cleric," was known "for his intimate involvement in multiple al-Qaida terrorist plots against U.S. citizens, including the 2009 Christmas Day bombing attempt in Detroit."[27] Both men were killed by drone strike in Yemen in September 2011.

Can we really say that these types of people deserve due process?

Much has been made about the killing of Abdulrahman Al-Aulaqi, Anwar Al-Aulaqi's sixteen-year-old son, outside a Yemeni café in October 2011. Once again, human rights groups and civil libertarians decried the use of extrajudicial killings. But former White House press secretary Robert Gibbs had a clear and relevant retort to critics that you can use as a template. When asked about the U.S. government having killed a minor without due process and without a trial, he said, "I would suggest that you [Abdulrahman Al-Aulaqi] should have a far more responsible father."[28]

This is the correct perspective. As Shakespeare so clearly put it, "The sins of the father are to be laid upon the children."[29] At sixteen, the boy certainly could have been expected to make better choices about his parents.

Before you become concerned that Americans will be unwilling to accept these restrictions, consider that when it comes to using methods such as drone strikes, American sympathies are overwhelmingly behind you. Some 58 percent of Americans surveyed in 2015 supported drone strikes "to target extremists."[30] This number climbed to 74 percent among Republican respondents, although more than half of the Democrats surveyed also offered their support. Only half (48 percent) of respondents were "very worried" about killing innocent civilians.[31] When it came to questions of legality, less than 30 percent were "very concerned" about whether drone strikes were being conducted legally. A full 17 percent weren't "concerned at all" about the legality of strikes, and another 16 percent stated they were "not too concerned."[32]

A Gallup poll from June 2002 found that nearly 80 percent of Americans were willing to give up some freedoms in the name of gaining security. Some 30 percent favored making it easier for officials to access private communications such as email.[33] When asked, in 2013, if the "federal government had computerized logs of your telephone calls or internet communications stored in a database that it uses to track terrorist activities, how concerned would you be that your privacy rights had been violated," 42 percent of those surveyed stated they were "not too concerned" or "not at all concerned."[34]

These trends have been persistent. A nationwide poll from 2015 found that just one-quarter of Americans said that protecting their rights and freedoms was more important than being kept secure.[35] Four in ten of those surveyed said that safety was more important than liberty.[36] When it came to such things as warrantless surveillance, more than half of those surveyed (56 percent) favored warrantless government collection and analysis of private data—even for U.S. citizens.[37] Analyzing responses along partisan lines, a whopping 78 percent of Republicans either favored or were indifferent to such activities. Some 68 percent of Democrats responded similarly.[38]

Whether it is the curtailment of free movement, free speech, or privacy, the public must be made to see these infringements as necessary to preserve safety and future freedoms. Never forget that fear is the foundation of your power over the public and required to get things done.[39]

The people of today may not be as free. But the people today are less important than the future generations at home and abroad who *may* enjoy the fruits of past and contemporary sacrifice. You must work to cultivate an attitude of fear, acquiescence, and proactive support among the populace: "Thank you, sir, may I have another?"

6

Embrace Top-Down Economic Planning

IF YOU HOPE to establish and maintain a global liberal order, you're going to need resources—a lot of them.

Free markets are wonderful for producing many goods and services, but unhampered private markets cannot be relied upon for matters of military production. As such, you must embrace state-led capitalism. A small group of state experts will need to carefully plan in conjunction with private firms. These private-public partnerships will form the basis for acquiring the goods and services you'll need to meet your foreign policy goals.

Ideally, these private-public partnerships will be voluntary on the part of firms. Indeed, many companies are happy to get in bed with Uncle Sam. Though not always a faithful lover, he has deep pockets.

Consider the major defense contractors—Lockheed Martin, Northrop Grumman, General Dynamics, and Boeing. This is but a sample of the harem of defense contractors and others who are more than happy to produce for your benefit. Contracts with the government mean a steady paycheck—and typically a large one. You need do little more than tell these contractors what you want and watch them produce it.

If individuals and firms fail to cooperate, however, this is not a problem. You have the ability to apply pressure through indirect means—for example, backroom cajoling and negative media coverage directed at uncooperative companies that are greedily driven by private profit opportunities. Consider, for instance, the rationing that occurred during World War II. "In addition to mobilizing industry, agriculture, manpower, and money, the federal government also had to manage the civilian economy to combat excessive price increases and to ensure an efficient and equitable distribution of consumer goods."[1] This involved setting price controls and output targets and imposing strict rationing.

Both citizens and businesses detested these policies, but most Americans would eventually agree that they were "a necessary part of the war effort."[2] Those who did not wish to comply—both producers and consumers—faced enormous pressure to capitulate. In addition to fines and other penalties for engaging in black-market activities, posters and other materials from the period likened failure to comply as tantamount to aiding the enemy:

> America has been attacked within its own borders! … Who are the Axis agents thus threatening our national safety? Saboteurs? Spies? Traitors? American quislings? No, they are none of these. … They are the buyers and sellers of "The Black Market." They are the bootleggers and purchasers of tires, gasoline, steel and irreplaceable critical materials, who evade the rationing laws and rules designed to assure the United States of sufficient military and industrial strength to win the battle for its existence and for the preservation of the rights of free men. Bootlegger and buyer are equally guilty. … They must be shown! They must learn that they are enemies of their nation![3]

Though some unscrupulous individuals and companies made use of black markets during the war, most complied with government requirements. Want to ignore or sidestep governmental guidelines during wartime? Then you're no better than any other Tojo flying a Zero into one of our warships.

If you cannot pressure domestic economic actors into compliance, force is always an option. The U.S. government has a long history of engaging in forced top-down economic planning when it comes to matters of war and national defense.

Consider, for example, the creation of the War Industries Board (WIB) during World War I. The Board was empowered, by order of President Wilson, to (1) create new facilities for producing resources for the government, (2) forcibly convert existing facilities to produce different outputs, (3) negotiate and set commodity prices, and (4) determine which materials to prioritize, among other things.[4]

Publisher Curtis N. Hitchcock described the WIB succinctly: "After a year of war the direction of industrial policy is placed in single hands, and a central planning board is established for dealing not only with the problems of production and purchase but with the whole attitude of the government toward the mobilization of business resources for the production of the war."[5] He described how the Board entirely jettisoned the coordination mechanisms of the free market and instead adopted command-and-control policies throughout the economy. Through willful manipulation, the WIB controlled the prices of everything from leather and lumber to steel, cotton, and glass.

> [The prices set by the WIB] tended to reduce to the minimum the great price stimulation which would have followed throwing on an expectant short market the vast early demands of the government for army supplies. In many instances this "pegging" of prices extended all the way down from the finished product to the raw material. In the case of shoes, for example … the price for tanned leather, hides,

thread, and almost every other component of the finished product had been arranged in advance by the government. ... The Secretary of War early suspended the statute making open-market bidding necessary.[6]

Hitchcock continued,

To avoid conflict between the [government] departments, when the prices were fixed the Board undertook the task of allocating to each department in the order it regarded as most important its share of the material in question, thus taking over in a voluntary but none the less effective way the control of the national supply of many commodities.[7]

The WIB is far from the only example of ways you can effectively engage in top-down government planning of the economy in the name of advancing liberal values.

During World War II, the First and Second War Powers Acts vastly increased the power of the executive. These bills would, according to Representative Hamilton Fish, "merely give the President ... powers that are necessary to win the war." Those powers were broad. According to the First War Powers Act,

the President is hereby authorized to make such redistribution of functions among executive agencies as he may deem necessary, including any functions, duties, and powers hitherto by law conferred upon any executive department, commission, bureau, agency, governmental corporation, commission, office, or officer, in such a manner as in his judgement shall seem best fitted to carry out the purposes of this title.[8]

The Second War Powers Act, enacted just a few months later, granted the president or his agents authority to acquire—voluntarily or not—private property for the war effort.

Following the conclusion of World War II, Congress approved the Defense Production Act (DPA) in 1950, recognizing that the battle against communism required the state to exercise the same powers it had during the earlier wars. The DPA provided "broad authority to the President to control national economic policy."[9]

> The DPA allowed the President, among other powers, to demand that manufacturers give priority to defense production, to requisition materials and property, to expand government and private defense production capacity, ration consumer goods, fix wage and price ceilings, force settlement of some labor disputes, control consumer credit and regulate real estate construction credit and loans, provide certain antitrust protections ... and establish a voluntary reserve of private sector executives who would be available for emergency federal employment.[10]

The DPA, though a product of the Cold War, is still an active policy today, though not all its provisions have been retained. It has been reauthorized more than fifty times, most recently in 2019.[11] The DPA will expire in September 2025, unless reauthorized by Congress. If history is any indication, however, you should not worry about losing this important tool.

Over time, the scope of the DPA has changed to include "national defense" in a broad sense. According to the Congressional Research Service, "the use of the DPA extends beyond shaping U.S. military preparedness and capabilities, as the authorities may also be used to enhance and support domestic preparedness, response, and recovery from hazards, terrorist attacks, and other national emergencies." The authors add, "Over the many reauthorizations and amendments to the DPA, Congress has gradually expanded the scope of the definition of defense, as recently as 2009. At that time, Congress included critical infrastructure assistance to any foreign nation and added homeland security to its definition."[12]

Under the current iteration of the DPA, you can supersede other contracts; what the government wants companies to make is paramount over their other obligations. This contracting authority is "routinely employed by the DOD," so utilizing this portion of the DPA will be unlikely to raise any eyebrows.[13] It has been used in recent years by other agencies as well, including the FBI, the Army Corps of Engineers, and the Federal Emergency Management Agency (FEMA).

Significantly, under the DPA the president maintains the authority of *priority performance* and *allocation*. The former requires individuals (businesses included) to accept government contracts. The latter, like the powers offered during World War II, allows the president to control the allocation of materials related to matters of national defense. It also empowers the president to issue loan guarantees and direct loans if needed. If the president determines such actions are necessary for national defense, he or she may require private facilities to install, modify, or expand production equipment. Also significantly, the president has the authority to delegate these powers. In 2012, for example, President Obama offered DPA authority to sixteen federal entities.[14]

The broad writing of the DPA and its frequent use (the DOD uses its DPA authorities to place some three hundred thousand orders a year) means that you will have the ability to take control of the economy in many situations—for the good of the people and for the success of your foreign policy.[15] During the COVID-19 pandemic, for instance, President Trump used his authority under the DPA some eighteen times. General Motors and General Electric were ordered to increase their production of ventilators—mechanical devices that "breathe" for patients who are unable to do so on their own. A variety of medical manufacturers were similarly ordered to ramp up production. Multinational conglomerate 3M, for example, was ordered to "acquire, from any appropriate subsidiary or affiliate of 3M company, the number of N-95 respirators that the Administrator determines to be appropriate."[16]

Not everyone will want to cooperate, of course. Undoubtedly, some companies will provide a variety of excuses to avoid your orders if they feel that their bottom line is threatened.

But you know better and, thankfully, have legal coercion on your side. When 3M was instructed to increase its domestic production and distribution of N-95 masks, for example, the company was also instructed to stop sending them abroad. In response, 3M stated that ceasing these exports (mostly to Canada and Latin America) would have "significant humanitarian implications … to healthcare workers" in those countries.[17] The company also argued that the order might backfire. "We are a critical supplier of respirators. … Ceasing all exports of respirators produced in the United States would likely cause other countries to retaliate and do the same, as some have already done. If that were to occur, the net number of respirators being made available in the United States would actually decrease."[18]

Pushback also existed in the case of the aforementioned ventilators. The U.S. government ordered some $3 billion in ventilators with the goal of creating a large national stockpile. These ventilators would save the lives of countless Americans. Some 140,000 machines were added—a tremendous success![19] A group of twenty-one "ventilator specialists" (mostly medical doctors), however, decided to diminish this undeniable triumph of government capabilities. They noted that there are different types of ventilators, not all of which are appropriate for treating COVID-19 patients. These "experts" claimed that only about half of the newly created machines had the ability to support severely ill COVID-19 patients.[20]

This is all nonsense, of course. Tell 3M, medical experts, and those like them to stick to what they know and leave the crafting of national security policies to the experts.

What does 3M know about the market in which it sells its products? What do medical doctors know about what types of materials are needed for the safety and security of the American and global public? Little to nothing. At best, they can see only a small part of the big picture, whereas you, the national security experts, see the global totality of the issue.

Moreover, they are self-interested capitalists; you are other-regarding public servants. If they don't want to do their part willingly, force them. It's really that simple.

President Trump wasn't the only one to invoke the DPA during the COVID-19 pandemic. President Biden also used the DPA. During his first day in office, President Biden issued an executive order directing the secretaries of defense, state, homeland security, and health and human services to review prices and availability of "critical materials, treatments and supplies needed to combat COVID-19." The order stated that, should these agencies identify any problems, they should "take appropriate action using all available legal authorities, including the Defense Production Act, to fill these shortfalls."[21]

Don't forget that the precedent is there for you to exercise your control of economic activity as you see fit.

Moreover, always remember that "national defense" is a broad and elastic concept. This means that you can define it as you like for your purposes. President Biden, for example, has issued no fewer than five executive orders, using the provisions in the DPA, to increase the production of solar panels. Are solar panels necessary for national defense? For pandemic response? Fortunately, the answer to these questions is up to those who get to define what counts as national defense—you! You have power and flexibility in controlling economic activity. Should you want to override private enterprise to advance a policy—whether defense-related or not—the DPA offers you a clear avenue.

Regardless of how the relationships between government and private industry are formed, it is critical to frame these partnerships in terms of ensuring safety, advancing freedom, enhancing liberty, and safeguarding free-market capitalism. In order to protect capitalism, we must suspend capitalism.

Take, for example, the "military-industrial complex" (MIC) or "military-industrial-congressional-complex" (MICC), the entanglement between private actors (contractors, defense firms) and the government's national security sector. In his farewell address, President Dwight D. Eisenhower coined the term. While he warned that "we must guard against the acquisition of unwarranted influence, whether sought or unsought," he also thought that such relationships were necessary.[22]

> We can no longer risk emergency improvisation of national defense; we have been compelled to create a permanent armaments industry of vast proportions. ... This conjunction of an immense military establishment and a large arms industry is new to the American experience. The total influence—economic, political, even spiritual—is felt in every city, every state house, every office of the Federal government. We recognize the imperative need for this development.[23]

Precisely. Markets are too unpredictable. Private individuals are too fickle and greedy. You must exert your authority to maintain economic control so as to bring order to chaos in the name of freedom and liberty.

Thankfully, you shouldn't need to do much to convince the public that eschewing the free market is necessary. Social scientists have provided you with the conceptual justification by labeling national defense as a "public good."[24] According to the logic of public goods, private markets are incapable of providing public goods in adequate quantities and qualities; citizens must, therefore, rely on the state to fill the gap. In most economics courses, national defense is *the* example of a public good.[25] One study, for example, found that 94 percent of economics textbooks cite national defense as an example of a good that must be provided by the state.[26]

The widespread acceptance of defense as a public good provides you with scientific cover to justify your policies at home and abroad.

Leverage these social scientists and their insights on national defense to further legitimize your national security efforts and to counter dissent. This is an easy task, because most social scientists already accept (and teach) that the state provision of defense and national security is a public good.

As one economist put it, "We are in a technological, economic, and arms-race competition with enemies with highly advanced tech capabilities ... and fewer scruples than we have regarding the use of government power. ... Everyone except a few die-hard ideologues and vested interests realizes that on some level by now."[27]

Indeed! Such extremist "die-hard ideologues" must be labeled as such and delegitimized for questioning the unquestionable necessity and supremacy of the U.S. national security state in centrally planning economic life.

This is especially important because, every once in a while, someone will try to reveal the true nature of this system to the public. For instance, the economist Robert Higgs has noted that the current U.S. military sector is an example of "military-economic fascism."[28] A fascist economic system is one whose essence is "nationalistic collectivism, the affirmation that the 'national interest' should take precedence over the rights of individuals."[29] In this system there is still private ownership of the means of production and of profit and loss.

But the state actively manages and partners with private economic actors for the national good.

Higgs is certainly correct in his characterization of the realities of the U.S. military sector. However, it is imperative that you do not let this rhetoric spread and take hold in the public psyche. If it does, it will threaten to undermine your noble goals.

Fortunately, it is easy to counter naysayers such as Higgs. Given the unquestioning attitude of most social scientists and the public that the state is necessary to provide national security, you can easily dismiss such dissent as the rantings of unpatriotic cranks who are not only members of the extreme ideological fringe but also naive regarding the realities of the world and the threats posed to the public.

A factor working in your favor is the sheer scale and scope of the U.S. national security sector, which includes a wide array of government and private actors. The sector is so expansive that even dedicated investigative journalists can't identify the specifics of its operations or who exactly has access to what.[30] You can be sure, therefore, that even the most interested ordinary citizen has practically no clue about what you and your colleagues are up to, the scale and scope of your operations, or the effects of the military sector on economic freedom.

In closing, we urge you to never forget to continually emphasize your unwavering commitment to free markets and capitalism, even when your actions are clearly at odds with the fundamental features of the market system and even as your actions march the country toward economic fascism in the name of combating that very ideology.[31]

Like other freedoms, economic freedom must be protected at *all* costs.

7

Loosen the Purse Strings

FREEDOM ISN'T FREE.

Maintaining global hegemony is expensive. This is not a new revelation, of course. Scholars have long noted that the U.S. government's many interventions, from the Spanish-American War to the Gulf War, have been costly.[1] For perspective, between fiscal years 2001 and 2022, post-9/11 conflicts are projected to cost $8 trillion when one considers estimated future obligations to veterans.[2]

While this amount may appear to be excessive, consider that, between 2018 and 2020 alone, the United States engaged in counterterrorism operations in at least eighty-five countries. So these funds are truly used to spread our ideals around the globe—not just to one or two countries.[3]

When you consider that the U.S. military is *the* harbinger and protector of liberalism at home and abroad, this price tag is a clear bargain!

It's important to note that the United States is not the only country responsible for picking up the tab for our nation's military operations. Other countries will also foot part of the bill. Consider that the governments of Japan, Germany, and South Korea compensate the U.S. government in a variety of ways for the privilege of hosting our troops in their countries.[4] Some of these payments are direct cash transfers to the U.S. government. Others take the form of tax waivers and investments in infrastructure (buildings, roads) associated with the military installations.[5]

Of course others can, and should, be urged to pay more. During his time in office, President Trump repeatedly stated that the U.S. government subsidizes our allies and that they, therefore, don't pay their fair share for security. He is certainly right; other governments should be willing to compensate the government for your costs and more.

The general lesson is that efforts should be made to have other governments pay as much as possible to alleviate the costs on American taxpayers. Foreigners may not know it, but they are privileged to have you intervening in their countries. It is only fair, then, for them to pay for the privilege, whether willingly or unwillingly.

War is no time for fiscal conservativism or "limited government." Payment terms can be reworked, and debts can be repaid later. The crisis is happening *now*, and it must be addressed *now*.

When a soldier sustains a major injury to his arm in battle, you apply a tourniquet. If you don't, he will die. You worry about his life first, his limb later. In the context of establishing global liberal order, fiscal concerns are like injured arms. The monetary costs of intervention (the limbs) are worth considering, but only after liberal peace (the body) has been established and firmly secured. Remember that the global military and security operations of the U.S. government are not an add-on, boutique program, but the very reason we have order and freedom in the first place. Without you and your top-down control, there would be nothing but chaos and anarchy. It is for this reason that you cannot let fiscal considerations get in your way.

Fortunately, you have multiple options for financing your mission of spreading peace and liberal values globally. The first of these options is direct taxation. In this case, the costs of your activities are passed immediately and directly to the citizenry through increased taxation. For example, shortly after the United States declared war on Germany in October 1917, Congress passed the War Revenue Act. The Act lowered the number of exemptions and greatly increased tax rates. Americans, who until just a few years prior likely paid *nothing* in income taxes, were required to open their pocketbooks to "provide means for paying for the war out of current taxes rather than through borrowings."[6]

The Act levied heavy taxes on beverages: $1.10 per gallon on distilled spirits, or $26.13 per gallon in 2023 dollars. Similarly high taxes were placed on tobacco products. Title VI of the War Revenue Act, titled "War Excise Taxes," levied taxes on "automobiles, musical instruments, jewelry, sporting goods and games, chewing gum and cameras … cosmetics, toilet articles and patent medicines … moving picture films … and motorboats and yachts."[7]

The tax had the desired effect. Tax revenues increased from $809 million in 1917 to $3.6 *billion* in 1918. One-third of the full cost of the war was paid for immediately through direct taxes.[8]

Taxes were raised to finance foreign military campaigns again during World War II.[9] President Roosevelt needed money to fund not only the war, but also his New Deal policies. In 1942, Congress passed a new Revenue Act, doubling the number of Americans required to pay income taxes.

Tax rates for the wealthiest individuals rose to 94 percent. Income taxes went from a "class tax" to a "mass tax" overnight.[10] Instead of paying taxes quarterly, citizens would instead pay their taxes on a rolling basis—with each paycheck, via withholding. The resulting spreading of tax payments not only diminished the "sticker shock" that many Americans (not used to paying income taxes) would feel but also ensured that the government received a steady stream of cash over time. Delivering his budget address to Congress in 1942, President Roosevelt made it clear why tax increases were necessary:

> In practical terms the Budget meets the challenge
> of the Axis powers. We must provide the funds to
> continue our role as the Arsenal of Democracy. …
> Powerful enemies must be outfought and outpro-

duced. ... We cannot outfight our enemies unless, at the same time, we outproduce our enemies. It is not enough to turn out just a few more planes, a few more tanks, a few more guns. ... We must outproduce them overwhelmingly, so that there can be no question of our ability to provide a crushing superiority of equipment in any theater of the world war.[11]

Called "the biggest piece of machinery ever designed to separate dollars from citizens," these new taxes had the desired effects.[12]

By 1945, some fifty million returns were filed, with taxpayers turning over $19 billion to government.[13] Paying income taxes became a point of patriotic pride. Even Disney's Donald Duck paid his fair share. In 1943, the cartoon fowl reminded Americans that taxes were "higher than ever before" due to "Hitler and Hirohito." They could "spend [their paychecks] for the Axis" or save and pay their taxes. "Every dollar you sock away for taxes is another dollar to sock the Axis. ... Taxes [will be used] to beat to earth the evil destroyer of freedom and peace. ... Taxes will keep democracy on the march," audiences were told.[14]

Irving Berlin, who during World War I wrote the still-popular "God Bless America," penned yet another tune in support of World War II. Instead of appealing to God, however, he appealed to his fellow taxpayers and their patriotism to defeat the Axis powers. The lyrics highlight the clear linkages between patriotism, taxation, and the success of Allied forces:

> *I paid my income tax today.*
> *I never felt so proud before,*
> *To be right there with the millions more*
> *Who paid their income tax today.*
> *I'm squared up with the U.S.A.*

See those bombers in the sky?
Rockefeller helped to build 'em, so did I.
I paid my income tax today.

I paid my income tax today.
A thousand planes to bomb Berlin.
They'll all be paid for, and I chipped in,
That certainly makes me feel okay.
Ten thousand more and that ain't hay!
We must pay for this war somehow,
Uncle Sam was worried but he isn't now,
I paid my income tax today.[15]

Unfortunately, today it is more difficult for you to effectively increase taxes to finance your benevolent hegemony. Celebrities are unlikely to come to your aid. If only Taylor Swift would create a ballad encouraging on-time tax payments or Lil Wayne would be willing to contribute a rap discouraging tax evasion to repent for his former transgressions against the Internal Revenue Service![16]

During the world wars, President Wilson and President Roosevelt had novelty, patriotism, and (at least at first) relatively low tax rates on their side. Prior to the wars, hardly any Americans had ever paid income taxes. Once the tax machinery had been established and people became accustomed to having their wealth transferred from their paychecks to government, officials understandably kept these higher taxes in place.

These higher taxes helped to pay for the Korean War in the 1950s. Even with tax cuts in the 1980s, revenue collections remained significant.

It may not be as easy to tap into new sources of tax revenue today. The Congressional Budget Office projects that government will have collected some $4.8 trillion in revenue in 2022, with revenue growth "mostly from large increases in the collection of individual income taxes."[17] Indeed, Americans today feel "overtaxed." The most recent polls find that a plurality of Americans—54 percent—report that their federal income taxes are "too high." Some 66 percent stated that they are either "somewhat" or "very dissatisfied" with how much Americans pay in federal income tax.[18]

This may not be the biggest issue with using taxes to pay for your policies, however. Taxation directly links the public to operations abroad—making them more likely to pay attention to your activities and the associated costs. Irving Berlin, in his aforementioned song, made this quite clear:

> *I never cared what Congress spent*
> *But now I'll watch over ev'ry cent,*
> *Examine ev'ry bill they pay,*
> *They'll have to let me have my say.*
> *I wrote the Treasury to go slow.*
> *Careful, Mr. Henry Junior, that's my dough!*
> *I paid my income tax today.*[19]

Take this as a warning when considering directly raising taxes to pay for your adventures abroad. Further increases in taxes could jeopardize support for your ambitions of global peace. Fortunately, you have other funding alternatives available.

One option is to reduce government spending in other areas to reallocate resources to war efforts. This suggestion is so absurd, however, it's hardly worth mentioning. Where would officials make spending cuts? Education? Health care? Infrastructure? Half-a-million-dollar studies on the "hydrodynamics of defecation"?[20] The $3.6 million National Security Agency parking garage?[21]

Of course not. Legitimate austerity has never been the answer for either major political party in Washington, DC, and it certainly isn't now given the urgency of the global situation. We must look at other alternatives, such as debt financing.

War bonds—used extensively during the world wars—are one option. However, since these bonds pay a relatively low rate of return, they can be a tough sell. It is typically necessary to appeal to patriotism or emotion to generate bond purchases, making it difficult to ensure a consistent stream of revenue. War bonds have the additional drawback of allowing the public to directly interact with the war. Like direct taxes, the immediate up-front cost of war bonds can cause the public to eventually withdraw their support from foreign policy.

Other forms of debt do not have this problem. In fact, many types of debt allow you to finance your activities abroad without the public knowing (or at least without the public caring).

This shift toward debt financing for war (and other government programs) began in earnest during the Vietnam War. Estimates of the final cost of the war vary. The Congressional Research Service, however, places the cost of the conflict at around $738 billion in 2011 dollars (or more than a trillion in 2023 dollars) with nearly all of the cost financed through debt.[22] This occurred for several reasons.

First, President Johnson was interested in not only funding the war against the Vietcong but also paying for his "Great Society" programs. The taxes required to cover both would have been politically unpalatable. Second, the administration wanted to avoid drawing any more negative attention to the war. With popular support abysmally low, avoiding further scrutiny was paramount. In her analysis of Vietnam War financing, political scientist Rosella Cappella summarized Johnson's strategy succinctly and highlighted the benefits of debt over direct taxation:

> The Johnson Administration did not want the public to interact with the war. Once taxes were raised, the administration believed that the public would become more aware of the course of the war, including unpopular casualties. Thus, Johnson deliberately chose to avoid public scrutiny by attempting

to shield the costs of the war through a war finance
policy that did not raise taxes.[23]

The trend established by Johnson continues. In the war on terror, nearly all operations have been financed through debt. The
Watson Institute for International and Public Affairs captures this
succinctly:

> Federal budgetary expenditures for the post-9/11
> wars include many expenses far beyond direct Con-
> gressional war appropriations. The approximately
> $2.3 trillion in Congressional appropriations
> through Fiscal Year 2022 for "Overseas Contin-
> gency Operations" … are just the tip of an iceberg.
> Other spending directly related to the War on Ter-
> ror includes additions to the Pentagon "base" bud-
> get, about $900 billion through FY2022. And while
> the U.S. paid for past wars by raising taxes and sell-
> ing war bonds, the current wars have been paid for
> almost entirely with borrowed money. … Through
> FY2022, the U.S. government owes over $1 trillion
> on interest on these wars. Even if war spending
> ceased immediately … spending on interest would
> continue to accrue, reaching at least several trillion
> dollars over the next few decades.[24]

Though this may sound politically precarious, issuing debt is an
excellent option that you should embrace. You avoid the immediate problem of the public interacting with the costs of your policies
through direct payments.

Not only that, both you and your constituents will likely be dead
and buried before the bills come due. Those who are very young—
and those who have not yet been born—are not resistant to contemporary government spending.

This is critical to realize because in post-1945 America, "the public has become almost uniformly unforgiving of fiscal sacrifice."[25] For this reason, you *must* leverage the advantages of the less-visible financing offered by debt to the greatest extent possible.

You will not have to answer questions about the debt—leave that to future policymakers to deal with.

Your political rivals may try to derail your financing plan under the guise of "accountability." In 2009, for instance, former congressional representative David R. Obey (D-WI) and several cosponsors introduced the Share the Sacrifice Act. The bill would have added a 1 percent surcharge on federal income taxes plus an additional percentage point for those owing more than $22,600.[26] As one commentator noted, the purpose of the bill was not actually to enact the tax; it was a political ploy. "It is doubtful [the bill will pass]. But that is not Obey's purpose. He will probably offer it as an amendment at some point just to have a vote. Republicans … will be forced to choose between continuing to fight a war that they started and still strongly support, or raising taxes, which every Republican would rather drink arsenic than do."[27]

You must not let such distractions divert you from the task at hand. Accusing dissenters of being unpatriotic and failing to support the troops is a good strategy to insulate yourself against such attacks.

And always remember, there are no fiscal conservatives in foxholes.

Some argue that debt financing is unfair. Is it right to burden our grandchildren and great-grandchildren with debt to finance the wars of today? In statements about the Share the Sacrifice Act, Representative Obey stated that it was wrong to burden future generations with our war spending.

While it is true that future generations will bear a financial cost, this objection to debt finance ignores an important fact. Future generations are the very people who will benefit most from the peace and prosperity brought about by the success of contemporary U.S. foreign policy. "The choice of how to finance a war is mainly a question of equity," according to the Congressional Research Service. "Financing through borrowing has been justified on the grounds that future generations benefit from the sacrifice that present generations make by fighting the war, and should therefore bear some of the cost of the war."[28]

Hear, hear!

We are the ones putting the skin in the game today. *We* are the ones doing the hard work today. The *least* that future generations, born and unborn, can do is to help pick up the tab.

Both the living and the unborn must sacrifice in the name of freedom, liberty, and democracy. We are one nation, a collective. And as a collective, individual people must sacrifice (and be sacrificed) for the common good; individualism depends on it.

The other alternative to fund your conflict is to turn on the printing press. One benefit of printing money is that it does not require consent from the public—only coordination with the central bank. The use of fiat—currency without any sort of commodity backing— means that the money supply can easily be increased in order to finance a war or to pay off war debts (i.e., monetize the debt).

Though wide-scale expansion of the currency undoubtedly leads to inflation, the public is largely ignorant of how inflation actually works. Further, the causal chain linking war to monetary expansion to inflation is a nuanced one, making it easy to shift the blame away from the original, genuine causes of higher prices to other factors such as greedy for-profit businesspeople or foreigners.

For example, responding to high rates of inflation following massive monetary expansion during the COVID-19 pandemic, officials were quick to pass the blame almost exclusively to supply chain problems and corporate greed—anything *but* the massive increases in the money supply. Maxine Waters, Democratic representative from California, for example, stated, "Right now, we're seeing big corporations take advantage of economic conditions and a lack of real competition to pass higher prices onto consumers simply because they can."[29] House Speaker Nancy Pelosi stated that higher gas prices were the result of "exploitation" on the part of oil companies.[30] (Offering a clinic in how to limit the conversation to prevent critical thinking by the citizenry, she skillfully provided no explanation for why gas prices also *fall*, as well as rise.)

Both debt financing and printing money are the most effective ways to fund your policies. They are both useful tools for engaging in the fiscal illusion necessary to conceal the costs of foreign policy at a time when generating and maintaining support for those policies is paramount.

The best part of these methods is that they hide the cost of the policies from the public—at least for now. The best thing about hidden costs is that they are hidden! Never forget the benefits of cost concealment, and use it to your full advantage whenever possible.

Kick the can down the road, good and hard! The future of freedom and democracy depends on it.

8

Silence Dissent

WE HAVE EMPHASIZED the crucial importance of crafting your narrative, capturing the media, and doing your best to ensure that the public is willing to provide financial and moral support to your noble global project.

Unfortunately, there will always be dissenters. Antiwar crusaders, antigovernment zealots, fiscal conservatives, and naive peacenik pacifists will inevitably try to disrupt and derail your plans. When this occurs, it is imperative that you act decisively.

Dissent is not acceptable when you are fighting for fundamental individual freedoms.

As the old saying goes, "A penny of prevention is worth a pound of cure." If you capture the media and control the narrative as we've suggested earlier, you can proactively mitigate dissent *before* it occurs. Other active steps to prevent dissent include suppressing or refusing to grant access to information, even when it is requested. In March 2023, for example, the Pentagon effectively blocked the Biden administration from sharing information with the International Criminal Court in The Hague about Russian war crimes in Ukraine.

Although this may seem an odd thing to deny, consider the following. If the court were to effectively investigate the Russian government for crimes, this could set a dangerous standard. "American military leaders oppose helping the court," writes journalist Charlie Savage, "because they fear setting a precedent that might help pave the way for it to prosecute Americans."[1]

Exactly! Global control requires significant power and the discretion for you to use that power as you see necessary. It is imperative, therefore, for you to push back against efforts to constrain your powers, both in the immediate term from dissenters and over the long term through legal and constitutional constraints.

You may also try to prevent dissent by cultivating the youngest Americans to be good patriots and by working to quiet or ostracize those who deviate. Consider, for example, the Pledge of Allegiance. *Every single day*, school children across forty-seven states are required to recite the pledge, stating their "allegiance to the flag of the United States of America, and to the republic for which it stands." Although the Supreme Court ruled in 1943 that children could not be required to say the pledge, reciting it remains a daily ritual for millions of American school children. And despite the Supreme Court's ruling, students have continued to suffer consequences for their lack of appropriate support for the state.

After opting out of an assignment requiring students to write the pledge in 2017, Texas student Mari Oliver stated that teachers "singled her out during the pledge, sent her to the principal's office, admonished her in class, and confiscated her phone."[2] That same year, high school student India Landry was expelled after multiple refusals to stand for the pledge. According to court filings, she'd refused to participate more than two hundred times! Eventually, the principal had enough of her disrespect and expelled her.[3]

In 2018, a middle school teacher in Colorado forced a student to his feet by his jacket and removed him from the room after he failed to say the pledge.[4] In 2022, a teacher in Florida berated a student who failed to stand for the pledge. In a video of the incident, the teacher confronts the student in front of the class. "Go back to your—where are you from? Mexico or Guatemala, where?" The student replied, "I was born here." "You were born here," the teacher responded, "and you won't stand up for the flag?!"[5]

The message here is clear and noble: *good* Americans support the flag and the government that flag represents; anti-American individuals do not. When your peers are all participating, you'd be smart to fall in line and participate, too.

The broader lesson is that social shaming and ostracism are powerful tools for making people fall in line and support your activities and goals.

Despite these preventive efforts, however, you must still be prepared to take further steps to squash discord. History again provides useful examples. Consider the use of the Alien and Sedition Acts. Passed in 1798 during a disagreement about what to do regarding the war between France and England, the Sedition Act provided a mechanism for the government to fine and imprison those speaking out against the United States—particularly the administration of President John Adams and the Federalists.

> If any person shall write, print, utter or publish, or shall cause or procure to be written, printed, uttered or published, or shall knowingly and willingly assist or aid in writing, printing, uttering or publishing any false, scandalous and malicious writing or writings against the government of the United States, or either house of the Congress ... or the President ... with intent to defame the said government ... or to bring them ... into contempt or disrepute; or to excite against them ... the hatred of the good people of the United States, or to stir up sedition within the United States, or to excite any unlawful combinations therein, for opposing or resisting any laws of the United States, or any act of the President of the United States, done in pur-

suance of any law … or to resist, oppose, or defeat
any such law or act, or to aid, encourage or abet any
hostile design of any foreign nation … shall be punished by a fine not exceeding two thousand dollars,
and by imprisonment not exceeding two years.[6]

For perspective, $2,000 in 1798 dollars is about $49,000 in 2023
dollars. The Founding Fathers were serious about sedition. Between
1798 and 1801, at least twenty-six individuals were prosecuted under the Sedition Act; "many were editors of Democratic-Republican
newspapers, and all opposed the Adams administration."[7]

The Alien and Sedition Acts all expired or had been repealed by
1802, save the Alien Enemies Act. This became particularly helpful
during World War I. In April 1917, Congress declared that any German
subject who had not been naturalized was an "alien enemy." These
enemies were banned from having a firearm or other weapons, barred
from coming within half a mile of any military installation or factory,
and required to register with the federal government. They were not
allowed to leave the United States without special court approval and
were forbidden to "write, print, or publish any attack or threats against
the Government or Congress of the United States, or either branch
thereof, *or against the measures or policy of the United States, or against
the person or property of any person in the military, naval, or civil service
of the United States … or the municipal governments therein.*"[8]

The purpose for this law was clear. In his war address to Congress,
President Wilson noted that, as is true today, the United States had
a responsibility to act and that "it will be all the easier to conduct
ourselves as belligerents in the high spirit of right and fairness because
we act without animus, not in enmity toward a people … but only in
armed opposition to an irresponsible government which has thrown
aside all considerations of humanity and of right." He highlighted
the need for quashing dissent: "If there should be any disloyalty, it
will be dealt with a firm hand of stern repression."[9]

Stern repression would indeed follow—as it should in the case of dissent. In June 1917, Congress enacted the Espionage Act, which made it a federal crime to interfere with the U.S. armed forces in any way. Anyone who spoke publicly against the war or draft could be investigated and prosecuted. The Act was a remarkably successful tool. Because of its vague wording, the government was effectually able to target anyone who opposed the war—from communists to pacifists. Some seventy-four newspapers were denied mailing privileges. Charles T. Schenck was tried and convicted under the Act for circulating a flyer opposing the draft.[10] The government was able to arrest ten people on actual charges of sabotage, which had the added benefit of quieting almost another 1,500 who, despite not being saboteurs, were nonetheless speaking out against U.S. foreign policy.[11] How dare they question government policy during a time of crisis?

The Sedition Act of 1918 further quieted dissent by curtailing free speech. Under the Act, whoever shall make statements to "incite subordination, disloyalty, mutiny, or refusal of duty … or shall willfully obstruct … the recruiting or enlistment service of the United States or … shall willfully utter, print, write, or publish any disloyal, profane, scurrilous, or abusive language about the form of government of the United States … or the military … or shall advocate, teach, defend, or suggest the doing of any of the acts or things in this section enumerated … shall be punished by a fine of not more than $10,000 [approximately $199,000 in 2023 dollars] or imprisonment for not more than 20 years, or both."

Many elements of the original Espionage Act remain in force today. While you may not think this of consequence, realize that the Act maintains remarkable contemporary relevance for you. This is particularly true in the case of whistleblowers. Whistleblowers—employees or former employees who reveal wrongdoings to Congress or the public—are dangerous.

As discussed earlier, one of the key advantages you possess is the presence of asymmetric information—differences in the information possessed by you and the information available to members of Congress and the public—in the national security state. By exposing information that would otherwise be kept secret, whistleblowers threaten this strategic advantage by closing information gaps between the national security elite and the public.[12]

This could spell disaster for your foreign policy plans.

Consider Daniel Ellsberg, the whistleblower responsible for releasing the Pentagon Papers, a Department of Defense document detailing how the government, specifically the Lyndon B. Johnson administration, had "systematically lied not only to the public but also to Congress" about Vietnam.[13] Ellsberg's release of the documents to the media was positively disastrous—and officials knew it. Speaking to President Nixon, White House Chief of Staff H. R. Haldeman said,

> What they really hurt [the papers released by Ellsberg] … is the government. What it says is— [Donald] Rumsfeld was making this point this morning—what—what it says is—to the ordinary guy … you can't trust the government; you can't believe what they say; and you can't rely on their judgment; and the—the implicit infallibility of presidents, which has been an accepted thing in America, is badly hurt by this, because it shows that people do things the president wants to do even though it's wrong, and the president can be wrong.[14]

Indeed, Ellsberg drew the ire of a cadre of officials and some members of the public. Speaking of Ellsberg, Henry Kissinger said, "That son-of-a-bitch . . . I am sure he has some more information—I would bet that he has more information that he's saving for the trial. Examples of American war crimes that triggered him into it [releasing the information]. ... It's the way he'd operate. ... Because he is a despicable bastard."[15]

Steps must be taken to avoid a repeat of the Pentagon Papers episode.

Fortunately, despite the existence of laws to protect whistleblowers (e.g., the Lloyd–La Follette Act of 1912, the Civil Service Reform Act of 1978, and the Whistleblower Protection Act of 1989), these laws present few practical obstacles for you. The Espionage Act, combined with a culture of secrecy grounded in patriotism, has proved to be remarkably effective at keeping would-be whistleblowers quiet and punishing those who decide to run their mouths or publish information. Ellsberg was prosecuted under the Act, at one point facing up to 115 years in prison. Though the charges against him were ultimately dismissed, he still paid dearly for his actions; the charges weren't dismissed until two years after his arrest.

The Espionage Act has proved to be especially helpful in dealing with whistleblowers in the war on terror. Under the Obama administration, a total of eight people were either charged or convicted under the act. Most recently, the Trump administration used the Act to charge former government contractor Reality L. Winner.[16]

Do not be shy about using, or threatening to use, the Espionage Act to make an example of whistleblowers and to discourage future "do-gooders" from spreading information you'd prefer to keep private. The costs to you are low, while the costs to the troublemakers are significant, not just in monetary terms but also professionally and personally.

You can also use means outside of formal laws to quiet dissenters. The case of Daniel Ellsberg is again illustrative. President Nixon, worried that Ellsberg might have information about clandestine bombings in Cambodia or Nixon's attempts to sabotage peace talks in 1967, approved the creation of a special unit to investigate Ellsberg. Known as the Plumbers (because it was their job to "fix leaks"), the group broke into the office of Ellsberg's psychiatrist to try to find information that might be used to blackmail or smear him.[17] Though the Plumbers were unsuccessful in finding useful information, their tactics serve as a shining example of how you can control those wishing to cause trouble, through direct and public attacks on their persons.

As another example, recall that after Army intelligence analyst Chelsea Manning released hundreds of thousands of documents related to U.S. activities in Iraq and Afghanistan, including communications about civilian casualties, her mental health and gender identity served as wonderful fodder to discredit her, in addition to her arrest and conviction under the Espionage Act. "The strategy is essentially 'nuts and sluts,'" she writes. "You paint the person who's done the government wrong as unstable or sexually deviant or ambiguous in some manner to delegitimize them."[18]

Personally discrediting dissenters is particularly helpful when you cannot use formal laws to discourage or punish dissension. For example, during the 1960s, COINTELPRO (short for "Counter-intelligence Program") worked to discredit and dismantle a cadre of subversive individuals and groups. Discussing the nature of the tactics employed, legal expert Brian Glick notes that the government worked "to sabotage progressive political activity" through "ongoing, country-wide ... covert action—infiltration, psychological warfare, legal harassment, and violence—against a broad range of domestic dissidents."[19] Targets included organizations such as the Black Panthers and various socialist groups, as well as specific individuals.

Boxer and antiwar activist Muhammad Ali was a target after joining the Nation of Islam and refusing orders to be inducted into the U.S. Army and serve in Vietnam.[20] Writer Ernest Hemingway, who had worked for the FBI while in Cuba, came under surveillance due to his ties to Cuban officials such as Fidel Castro. The FBI conducted some fifty searches of Hemingway's property. Hemingway knew it, too. Recalling the period just before his suicide, his closest friends state that he was on high alert and concerned about being followed.[21]

Remember, *no one* is off-limits when it comes to fulfilling your plan. Even former friends and aides aren't above suspicion.

Dr. Martin Luther King Jr. was also a target of COINTELPRO. Combining both civil rights and antiwar rhetoric in his speeches, King was considered particularly dangerous, and rightfully so. He urged his fellow Americans not to offer their support for the U.S. government's war in Vietnam.

"I come to this platform to make a passionate plea to my beloved nation," he said in 1967. "I wish not to speak with Hanoi ... but rather to my fellow Americans." He went on to outline seven reasons for his opposition to the war, stating that "it should be incandescently clear that no one who has any concern for the integrity and life of America today can ignore the present war. ... It [America's soul] can never be saved so long as it destroys the deepest hopes of men the world over."[22]

Such rhetoric could not stand, of course.

The FBI began heavily surveilling King in 1962, tapping his home and office phone lines and placing bugs in his hotel rooms. They quickly learned of his extramarital affairs. Despite the FBI's attempts to offer this information to the press, the story generated little attention.

So, the FBI penned an anonymous letter to King, urging him to commit suicide. The letter, which can serve as a sort of template for your future efforts, reads in part:

> King,
>
> In view of your low grade, abnormal personal behavoir [*sic*] I will not dignify your name with either a Mr. or a Reverend or a Dr. . . ..
>
> Look into your heart. You know you are a complete fraud and a great liability to all of us Negroes. ... You are a colossal fraud and an evil. ... You could

not believe in God and act as you do. Clearly you don't believe in any personal moral principles. …

No person can overcome facts, not even a fraud like yourself. Lend your sexually psychotic ear to the enclosure. You will find yourself and in all your dirt, filth, evil and moronic talk exposed on the record for all time. … It is all there on the record, your sexual orgies. Listen to yourself you filthy, abnormal animal. …

King, there is only one thing left for you to do. You know what it is. … You are done. There is but one way out for you. You better take it.[23]

Whistleblowers and activists will not be the only ones you will need to quiet. Celebrities present another annoyance. When singers, actors, and other entertainers decide to use their fame for political opposition, you will need to act quickly and firmly.

As an example, recall that in 1972 actress Jane Fonda visited North Vietnam. A major figure in the antiwar movement, Fonda visited Hanoi and surrounding areas, claiming that the U.S. government was attacking nonmilitary targets. She spoke several times on North Vietnamese radio, imploring U.S. pilots to stop their bombing runs. She referred to South Vietnamese troops as "cannon fodder for U.S. imperialism."[24]

When she returned from her trip, she toured college campuses, spreading her antiwar message. These actions earned her the nickname "Hanoi Jane." The backlash to her activism was decisive. Some veteran groups and policymakers called for trying her for treason or sedition under the Logan Act (a federal law passed in 1799 making it a crime for an unauthorized American citizen to negotiate with a foreign government).[25] The barrage of criticism has lasted more than forty years. As recently as 2021, officials have continued to assail

Fonda's antiwar actions. Former Trump advisor Stephen Miller told Fox News that "by any definition, what [Fonda] did in the Vietnam War was treason."[26] The message is clear—criticism of U.S. policy, and particularly criticism of the U.S. military—is off-limits and will be met with harsh and long-lasting retribution.

Fonda certainly isn't the only celebrity critic of U.S. foreign policy. Shortly before the launch of the war in Iraq in 2003, Natalie Maines, Texas native and lead singer of the Dixie Chicks (now the Chicks), decided to express her opposition to the war at a concert: "Just so you know, we're on the good side with y'all. We do not want this war, this violence, and we're ashamed that the president of the United States is from Texas."[27]

In this case, officials had to do very little to actively silence Maines and her bandmates. As discussed earlier, the Iraq War narrative had been sold so well to the American public, and the media had been so well controlled, that fervent support was already high. The public and the media were happy to silence the Dixie Chicks on the government's behalf!

Maines and her bandmates were dubbed "Saddam's Angels," and radio stations quickly pulled their music, refusing to play it on air. Two disc jockeys in Colorado were suspended after defying their general manager's order to stop playing Dixie Chicks songs.[28] Local radio stations and other groups organized events to burn or destroy the band's CDs. As journalist Laura Snapes notes, this had the effect of "hobbling their career overnight. They would release one more album, in 2006, their last for 14 years."[29] Even though Maines quickly issued an apology, it was too little, too late.

The moral for celebrities? Shut up and sing. Shut up and act. Shut up and entertain. Leave the foreign policy to the experts.

And because celebrities are public-facing, a well-executed campaign to squash dissent will quickly create common knowledge among members of the public, making it more likely that they remain in line with your program.

In addition to whistleblowers, activists, and celebrities, there is one other group you must take special care to quiet. Even when you are able to successfully capture the media, there will undoubtedly be problematic journalists, determined to derail your efforts. We have already discussed how you should provide friendly journalists with exclusive access through embedding. But you will still need to actively push back against unfriendly or hostile members of the press. More contemporary efforts at silencing journalists often enlist cooperation from social media giants and other hosting platforms.

In 2010, for instance, financial technology company PayPal suspended payments to WikiLeaks—a nongovernmental organization infamous for publishing news leaks and classified documents from anonymous sources—after receiving pressure from the State Department.[30] More recently, PayPal banned two independent media platforms (MintPress and Consortium) from receiving payments and froze their assets in 2022. Though PayPal has not stated why the platforms were banned, journalist David Morris notes that both media outlets "recently published articles that challenge neat narratives about Russia's invasion of Ukraine. To be clear, [the outlets] are not peddling Russian apologism. ... Rather, they are antiwar." He continues, "The incident is worrying because the PayPal bans appear to be coordinated. That strongly suggests pressure from government authorities."[31]

And problematic media outlets like these two should be worried! You have the power to bend the will of private media and financial firms to serve your purposes. You would be wise to use it.

PayPal isn't the only organization working to quash journalists at the behest of authorities. In March 2023, the U.S. House of Representatives released a report titled "The Weaponization of the Federal Trade Commission" (FTC), which discusses the FTC's interactions with Twitter (now X) following its purchase by Elon Musk. The report describes how, following Musk's takeover, the company allowed journalists to see a cache of internal documents concerning its staff's interactions with government:

Twitter allowed ... journalists, as part of their reporting on government censorship by proxy, to review internal communications and correspondence between Twitter employees and federal agencies, including the Federal Bureau of Investigation. The journalists' reporting did *not* concern private user data. ... Quite the opposite, the reporting ... concerned content that users attempted to publicly share but that the government had pressured Twitter to restrict. ... The FTC's first demand in its letter ... demanded that Twitter "[i]dentify all journalists and other members of the media to whom" Twitter has granted access since Musk bought the company. The FTC even named some of the specific journalists. ... The FTC also demanded to know any "other members of the media to whom You have granted any type of access to the Company's internal communications" for any reason whatsoever.[32]

Of course, it is less than ideal that this information has become public. But those who discuss this information and critique the government can easily be dismissed as "unpatriotic" and "conspiracy theorists."

Taken together, what all this makes clear is that you have many diverse tools available to you to quash dissent. With sufficient planning, you can smother opposition in its cradle. When you cannot, take heart. Whether journalists, activists, or celebrities are the ones causing you trouble, remember that you have options. From using the law to prosecute those who dare expose the skeletons in Uncle Sam's closet, to surveillance, to using private companies to limit the reach or financial assets of an individual or a group, to public shaming by labeling people as unpatriotic traitors, rest assured that you can compel malcontents into silence—one way or another.

In the 1992 movie *A Few Good Men*, Colonel Nathan Jessup (played by Jack Nicholson) famously shouted, *"You can't handle the truth!"* In foreign affairs, truer words have never been spoken. Never forget that the general public cannot handle the truth about the messiness and complexity of foreign affairs. Democratic self-governance may apply to some domestic affairs, but certainly not to international affairs, which are beyond the grasp of the ordinary person.

It is your duty to silence dissent for people's own good and for the good of the republic and all it stands for—liberty, freedom of speech, freedom of religion, freedom of assembly, and due process of law.

9

Ignore International Law

RUNNING THE WORLD'S most powerful military on a global scale requires a certain mindset. You must maintain your confidence in your ability to solve complex problems in other societies. You must not forget that you possess superior knowledge and that your preferences are better than the preferences of the subjects of your intervention.[1]

It should be obvious at this point that authorities in this area undoubtedly enjoy knowledge that the public and those abroad simply do not have. It's *abundantly* clear that the preferences of the national security elite are better than those of individuals outside the United States. The fact that people in other societies seemingly enjoy living in squalor, with constant conflict, and under brutal dictatorships or other patently backward regimes illustrates that they're simply not capable of establishing better institutions on their own.

This fact has long been recognized. For example, Major General William R. Shafter, Union Army officer and commander in Cuba during the Spanish-American War, when discussing how a liberated Cuba would fare under self-governance, said: "Self-government! Why those people are no more fit for self-government than gunpowder is fit for hell."[2]

He wasn't wrong. After all, look at Cuba today. The manifestation of Cuban self-government came in 1959, when Fidel Castro took over the government. After *decades* of bad policy, *decades* of being little more than a pawn for the Soviets, the Cuban people did nothing to remedy their own situation. They continue to do nothing. Writing in 2023, one commentator noted, "Cuba's communist regime is at its weakest point in decades. The island's economic woes, brain drain, regime persecution of dissidents, and decaying state institutions are all exacting a high toll, but given authorities' repressive hold on society, it is unlikely that change is on the horizon."[3]

The Cuban people can't overthrow a crumbling regime keeping them in poverty, and yet they are expected to govern themselves? They've shown themselves to be incapable.

More recently, similar logic has been used in the context of the ongoing Israel-Palestine conflict. Some have argued that "the people" of Gaza are generally responsible for Hamas (and indirectly for the brutal atrocities committed by Hamas on October 7, 2023) because they voted for Hamas in the 2006 elections and because they haven't since removed Hamas from power.[4] This reasoning has been used by some to justify the Israeli government's current actions in Gaza but also as a foundation for long-term occupation.

The broader lesson is this: When contending with populations who aren't suited for self-determination, you must not let those populations and others deter you with their talk of "sovereignty" and "international law." In fact, it may be necessary to ignore these ideas entirely, for many societies will fall into chaos if you don't. Consider, for instance, any number of former colonies in Africa, Asia, and the Middle East. Time and again these countries have shown what happens when we fail to intervene or hold back due to "legal considerations" or because of concern about what the international community will think.

The genocide in Rwanda in 1994 and subsequent civil conflict in neighboring Zaire (now the Democratic Republic of the Congo) offer a relevant example. Allied Joint Force Command advisor Mel McNulty summarized the lead-up to Western involvement in these engagements as follows: "[The portrayal] of these conflicts as ethnically-driven facilitated Western interventionary responses, the rationale for which may be summarized as 'they are mad, we are sane, we must save them from themselves.' ... New outbreaks of conflict were interpreted as ethnically-driven ... inviting a further interventionary response."[5]

Sanity was clearly needed in Rwanda, but the United States failed to intervene militarily, having just received a barrage of criticism over leading the United Nations' operations in Somalia.[6] As a result, between five hundred thousand and one million Rwandans unnecessarily died at the hands of their countrymen. His failure to intervene remains a black mark on Bill Clinton's presidency.[7]

The reality is that men such as Shafter and McNulty hit the nail on the head. Some people are not fit to govern themselves. The people of some nations behave as though they are insane. They need a helping hand.

That hand must often be firm, ignoring the rules created by those who have no business telling us how to conduct our affairs. This view of the world has a long tradition in America. In 1899, Rudyard Kipling published his famous poem "The White Man's Burden."[8] There, Kipling urged the American government to establish colonial control over the Filipino people, who were neither ready for nor capable of freedom or self-governance. In a similar spirit, historian Niall Ferguson recently reminds us that "in many cases of economic 'backwardness,' a liberal empire can do better than a nation-state. . . . A country like—to take just one example—Liberia would benefit immeasurably from something like an American colonial administration."[9]

The biblical book of Proverbs reminds us, "Whoever spares the rod hates their children, but the one who loves their children is careful to discipline them."[10] Without other imperial or liberal powers to keep the more backward nations—such as Cuba, Rwanda, Somalia, and so on—in line, it is your job as agents of the benevolent hegemon to wield the rod.

Indeed, one of the key mentalities necessary for successful interventions abroad is comfort with employing various tactics, including uncomfortable ones. Success in foreign intervention "requires a willingness to use various techniques—monitoring, curfews, segregation, bribery, censorship, suppression, imprisonment, torture, violence, and so on—to control those who resist foreign governments."[11]

It's an unfortunate reality that such illiberal tools must often be considered and employed for the establishment and maintenance of liberal ends. But rest assured that nothing you will do to these nations is worse than what they will do to themselves absent your intervention. You cannot, as the old saying goes, turn a sow's ear into a silk purse. You cannot turn a savage into a sophisticate.

Fortunately, the constraints facing the United States in foreign engagements are weaker than those faced domestically. At home, the Constitution poses a barrier that must be circumvented through secrecy and manipulation. As the Constitution does not follow the flag abroad, those involved in national security matters have significant additional leeway.[12] They have much greater scope to do as they please without the extra effort involved in paying lip service to constitutional constraints while figuring out how to avoid them.

One of the best examples of the importance of ignoring international laws relates to the use of torture. Torture is expressly prohibited by the 1948 Universal Declaration of Human Rights and the 1966 International Covenant on Civil and Political Rights. It is considered a war crime under the 1949 Geneva Conventions. The United States adopted the United Nations Convention Against Torture (UNCAT) in the 1980s, making it U.S. federal law under the Supremacy Clause of the Constitution.

Nevertheless, you can feel free to ignore these rules in the service of achieving your goals. Who is going to enforce them? What are they going to do to you, the elites, in the most powerful nation on earth?

Anyway, what exactly is "torture"? Fortunately, that's for you to decide.

Keep in mind that the U.S. government has a long history of using torture as a means of spreading liberty and advancing foreign policy goals, with few repercussions. As in other instances in this guide, history proves illustrative. In 1898, the U.S. government assisted the Philippines in ending several hundred years of Spanish rule. The Filipinos assumed (foolishly, as Kipling recognized in his poem) that they would be independent after the Spanish made their exit. The U.S. government was forced to annex the island chain, leading to a fierce decade-long insurgency. Quashing this insurgency necessitated the use of several tactics, including enhanced surveillance and torture. One of these methods, the "water cure," saw extensive use.[13] A form of "pumping," whereby the stomach and intestines are forcibly filled with water, the water cure proved a popular method of extracting information from and imposing punishment on those who refused to accept U.S. authority. One witness described the water cure in the following terms:

> The water cure is pure hell. The native is thrown upon the ground, and, while his legs and arms are pinioned, his head is raised partially. … [T]he water is poured in, and swallow the poor wretch must or strangle. A gallon of water is much, but it is followed by a second and a third. By this time the victim is certain his body is about to burst. But he is mistaken, for a fourth and even fifth gallon are poured in. … While in this condition, speech is impossible; and so the water must be squeezed out of him. This is sometimes allowed to occur naturally, but is sometimes hastened by pressure and "sometimes we jump on them to get it out quick."[14]

The discomfort of a few for the greater good is necessary when you are seeking to spread liberal values, especially when you are dealing with "humans" who are incapable of self-governance or of understanding what freedom requires.

More contemporary examples of torture include the use of "enhanced interrogation" techniques during the war on terror, particularly in Iraq and Afghanistan. In Abu Ghraib, for instance, members of the U.S. military subjected detainees to a number of tortures, including sleep deprivation and stress positions. Individuals held at Guantanamo Bay, Cuba, were subjected to waterboarding and extreme isolation.[15]

Over the years, it has become progressively important to employ the use of "clean" torture techniques in the vast majority of cases.[16] These techniques induce the desired agony but, when done correctly, leave no lasting marks. This is important because, even though violating international laws regarding torture is necessary for spreading liberalism, torture is *incredibly* unpopular and a bad look for those proclaiming to be the standard-bearers of liberal values.

As we've made clear, it is imperative to mitigate anything that may diminish support for your policies or your goals. Clean torture is much more difficult to identify after the fact, which reduces the likelihood that there will be any consequences for those employing these techniques. Fortunately for you, clean torture has already become established as the U.S.-preferred technique.

Political scientist Darius Rejali highlights how the U.S. government has effectively shifted to these techniques and the benefits to using them:

> White House lawyers itemized techniques that would not, in their view, constitute torture. ... Phillip Heymann, a former deputy attorney general under the Clinton administration, proposed that American interrogators use "highly coercive interrogation methods." Heymann ... characterized these methods as techniques that fall midway

> between torture ... and non-coercive interroga-
> tion. ... Whether they are called ... "enhanced in-
> terrogation" or "highly coercive interrogation," all
> this is simply another way of saying they are clean,
> and therefore, misleading, techniques to those who
> observe them at a distance. Their main value is that
> they are gray tortures that are hard to condemn. In
> the absence of visible marks, how can anyone tell
> how much pain prisoners are in?[17]

Ivory-tower academics aren't the only ones to explicitly recognize these benefits. As one Department of Defense official put it, "If you don't make them bleed, they can't prosecute you for it."[18] Signs hung in Camp Nama, a former U.S. military base in Baghdad, reminded interrogators, "NO BLOOD, NO FOUL."[19]

To this effect, we urge you to consider employing any of the following clean torture techniques, each of which has been used by the U.S. government in the past:

Electrocution. This was a favorite during Vietnam. Interrogators often modified field telephones in order to shock the "interviewee" and obtain information. As former sergeant D. J. Lewis put it, "It was not uncommon ... to rig up a field telephone, and put one [wire] around a finger and the other around the scrotum and start cranking."[20]

Sleep deprivation. In the war on terror, the U.S. military authorized sleep deprivation for detainees for up to three full days. This technique is often paired with the use of stress positions to maximize effectiveness.[21]

Audio torture. This torture is divinely inspired, so you can be sure it is worthy! Discussing this technique, retired Lieutenant Colonel Dan Kuehl noted that in the Old Testament, Joshua's army used trumpets to break the spirit of those in the city of Jericho. Take a page from the Good Book and the war on terror and consider putting on blast "I Love You" by Barney the Purple Dinosaur, the jingle for Meow Mix cat food, or Metallica's "Enter Sandman," and watch your detainees crack like walnuts.[22] An added benefit is that this form of torture, considered "mild," usually induces chuckles from spectators as opposed to concern. "We've been punishing … our wives … with this music forever. Why should the Iraqis be any different?" quipped Metallica front man James Hetfield when discussing the use of his music as torture in Iraq.[23]

Exhaustion exercises. From continuous running and repeated squats, to handcuffed push-ups and walking on all fours like a dog, prisoners in Abu Ghraib and elsewhere have been subjected to forced exercise.[24] This "torture" is basically a tough day at the gym. While their muscles are screaming with pain, however, their mouths will be screaming out valuable intelligence. An added benefit for these detainees is the health benefits from the exercise routine associated with this technique.

Waterboarding. Used throughout the war on terror, waterboarding was explicitly authorized by the Justice Department following 9/11. Waterboarding takes several forms, often involving strapping an individual to a board and then pushing them under water. Other variants include wrapping a prisoner's head in cellophane and pouring water over the head to simulate drowning.[25] No one *actually* drowns; it's really no different from being dunked underwater in the swimming pool, as kids do all the time without anyone complaining.

Note that this is only a subset of the options available to you—not a comprehensive list. Conducting mock executions, administering "moral torture" (such as forcing Muslim detainees to be in the presence of dogs or to wear dog collars, or desecrating holy books), stripping detainees of their clothing, and threatening to rape detainees (or their wives and daughters) are all techniques that have also been used and may prove helpful to you.

You should also feel free to innovate and develop new techniques that will aid your colleagues in our shared goals of promoting freedom. The introduction of new techniques not only will help to spread and protect liberty but may even earn you a promotion as well!

In addition to acquiring useful information, remember that torture may also prove to be an effective means of broader social control to ensure that people in the wider society also fall in line. One commentator, discussing her experience of torture in Guatemala, stated this effect clearly.

> It is often assumed that torture is conducted for the purpose of gaining information. It is much more often intended to threaten the population into silence and submission. What I was to endure was a message, a warning to others—not to oppose, to remain silent, and to yield to power without question. … I was a message board upon which those in power would write a warning … [stop resisting] or be prepared to face the full force of the state.[26]

In addition to torture, you may also be compelled to use other, more intensive tools of war. For example, as part of the battles in Fallujah during the Iraq War, U.S. forces used chemical weapons containing white phosphorus. Other assaults necessitated the use of napalm. These chemical agents, which can cause serious burns or death, were necessary to secure victory in the U.S. government's attempt to export democracy by establishing a sustainable liberal regime in Iraq.[27]

As in other cases, you must take great care to minimize or conceal these activities, particularly from your domestic population. If the American public were to learn, for example, that the thermobaric weapons used by Marines in Fallujah, Iraq, are the likely cause of serious birth defects—in infants who have "organs spilling out of their abdomens … with their legs fused together like mermaids' tails" or who have "emerged [from their mother's wombs] gasping, unsuccessfully, for air"—this news may be poorly received.[28] Dying newborns are a *tough* sell.

When your actions do come to light, proactive messaging will be required to paint critics as romantic utopians and peaceniks who do not understand the realities of the world or warfare and who are sympathetic to our enemies and their illiberal values.

An alternative way to get around international law is to have others violate it for you. Consider, for example, the School of the Americas (SOA), rebranded in 2001 as the Western Hemisphere Institute for Security Cooperation (WHINSEC). Located in Fort Benning, Georgia, the school was founded in 1946 at the outset of the Cold War as the "Latin American Training Center." The school has trained tens of thousands of Latin American military members in dozens of courses, from basic patrolling techniques to advanced helicopter flight instruction. With a mission to "provide doctrinally sound, relevant military education and training to the nations of Latin America; to promote democratic values and respect for human rights; and to foster cooperation among multinational military forces," the school has been an invaluable asset in strengthening regimes friendly to the United States and—significantly—hostile to (or at least unfavorable to) communism and the former Soviet Union.[29]

While the school provides training in human rights "both formally—in classroom instruction … and informally—through exposure to American institutions," graduates have also been exposed to the *real* tools they need to get the job done.[30] The school's training manuals on terrorism and urban guerrilla warfare, interrogation, and combating communist guerrillas, for instance, suggest "the use of executions, beatings, extortion, and blackmail to fight insurgencies."[31]

When people pointed out that the manuals were instructing readers to engage in human rights violations, this moral hand-wringing was met with a masterful deflection of criticism. According to the Department of Defense, those instructions in the manuals were the work of "misguided junior officers working from outdated intelligence materials." Moreover, "[those creating the manuals] simply assumed that U.S. laws against assassination, beatings, and blackmail applied only to U.S. citizens and thus were not applicable to the training of foreign military officers."[32]

Over the nearly seventy years the school has operated, it has trained some of the most well-known leaders in Latin American history. General Manuel Antonio Noriega and General Omar Torrijos of Panama and Major General Guillermo "Bombita" ("little bomb") Rodríguez of Ecuador all graduated from the school. Lieutenant Generals Roberto Viola and Leopoldo Galtieri of Argentina and General Hugo Banzer Suárez and Major General Guido Hernán Vildoso Calderón of Bolivia were all distinguished graduates. While these men may have failed to absorb the "respect for human rights" portion of SOA programming (to be fair, that takes up only eight hours of their training), they clearly internalized the other key lessons. Each of these men led a military dictatorship in Latin America and has been accused of a number of human rights violations. Under Hugo Banzer's regime, for example, Bolivian leftists were "disappeared"—murdered or deported.[33] Galtieri was charged in the torture, abduction, and murder of some twenty dissidents in Argentina.[34]

But none of these countries turned communist, and most of them, to this day, remain friendly to the interests of the U.S. government—which is really the whole point.

You won't just want to train them, however. You will want to ensure that they are well equipped to leverage the skills you have taught. Fortunately, you have a long history to draw from when it comes to arming foreign agents. The U.S. government remains the world's largest arms dealer, despite signing the Arms Trade Treaty (ATT) in 2013, which expressly prohibits the transfer of any weapons that could be used to "commit acts of genocide, crimes against humanity or war crimes" and requires states to "deny an export if there is an 'overriding risk' that weapons will … undermine international humanitarian/human rights law … [or promote] organized crime, and gender-based violence.[35] Consider, for instance, that the U.S. government supplied 39 percent of global arms exports between 2017 and 2021, far outpacing its closest rivals.[36] Over that same period, Russia (with a 19% share of global arms exports) was second, followed by the governments of France (11%), China (4.6%), and Germany (4.5%).[37] In terms of recipients, governments in the Middle East received 43 percent of the U.S. government's total arms exports over the 2017–2021 period; the government of Saudi Arabia was the largest single recipient of U.S. government-provided weapons, accounting for 23 percent of total U.S. arms exports.[38]

Though these weapons may be used in ways you wouldn't prefer, once they leave your shores, that really isn't your concern. The immediate observable benefits to you and your goals need to be your main focus.

Consider, for example, Cold War weapons transfers to the government of El Salvador. Engaged in a brutal civil war for some twelve years (1979–1992), the Salvadorean government was an ally of the United States when Cold War concerns were paramount. The United States was not interested in letting El Salvador fall to leftist guerrilla groups and took active steps to prevent this from occurring. Although President Carter briefly cut off funding to El Salvador following the rape and murder of three American nuns and a U.S. missionary in the country, President Reagan reversed course, understanding the bigger picture: another Latin American country could not turn toward the Soviets. The administration poured millions of dollars into the country along with other military aid.

Indeed, some of this aid was used in a way that some have questioned. Consider, for example, the "massacre" at El Mozote in December 1981. The U.S.-trained Atlácatl Battalion, led by SOA graduate José Domingo Monterrosa Barrios, entered the village of El Mozote and quickly separated the men, women, and children. After killing the men, the soldiers raped and murdered the women and older girls before turning their attention to the children, slitting their throats or hanging them from trees. (The youngest victim was reportedly two years old, and most victims were minors.)[39] A later analysis would reveal that, unfortunately, U.S. materials were used. Archaeologist Douglas Scott writes,

> With one exception, the [shell] cases [at El Mozote] were all fired in United States M-16 military rifles. The single exception is a 7.62 mm NATO [North Atlantic Treaty Organization] case possibly fired in a United States M-14 rifle. … All cartridge cases, including the 7.62 mm NATO case, were head-stamped "L C," which indicates they were manufactured for the United States Government at Lake City Ordnance Plant located near Independence, Missouri.[40]

Despite all of this evidence, few people know about the connection between this event and U.S. government-provided weapons. The transfer of arms from one group to another provides a layer of protection. Once the weapons leave our hands, after all, how can we be held accountable? If someone steals your car and robs a bank, are you as guilty as the robber? Of course not. Besides, it's possible that those village children were planning a communist revolution!

In other cases, weapons transferred to target groups might be lost in the chaos of conflict and bureaucratic red tape. According to one study, the U.S. government transferred 1.45 million firearms to people in Afghanistan and Iraq in efforts to combat insurgents and establish police and defense forces.[41] Although the exact number is unknown, it is estimated that hundreds of thousands of firearms were, and still are, unaccounted for.[42] When these types of situations occur, it is best to highlight the complexity and pace of the situation and to highlight the costs of nonaction: the enemy will win absent the fast and decisive transfer of weapons and military equipment.

Indeed, a common theme running through this manual is that framing things in terms of fear-inducing counterfactuals that are broad and unfalsifiable is key to success. Never forget this general operating principle.

Though perhaps regrettable, the misuse of arms must be understood within a broader context. As we wrote earlier, you must be prepared to "crack some eggs" for the greater global good.

Sometimes, achieving a liberal order requires you to make deals with the devil.

Remember that the sins committed by others are not your sins. Once you've achieved your goals, all will be forgotten and forgiven, and these sacrifices will be viewed for what they are—necessary steps on the path to freedom.

10

Do Not Accept Failure

WHILE WE ALL know that perfection is not always possible, you should make every effort to always project your impeccability. Even then, the reality is that there will be times when things do not go according to plan. Unintended or perverse outcomes are bound to arise, given the sheer number and scale of the interventions that the U.S. government undertakes.

As discussed earlier, those who would fight against the U.S. government's project of global liberalism, both at home and abroad, will seek to capitalize on every misstep and use it as an opportunity to discourage public support. Worse, they may convince members of the public that your policies should be reversed or abandoned completely. Failure—or the perception of failure—gives the impression that the United States is unnecessary or even counterproductive in establishing and maintaining a global liberal order.

We have provided advice on ways to mitigate failures and mistakes: ignore them, conceal them, and use the media and other tools to minimize or even capitalize on them via spin. As we noted, the media is often uncritical and faces strong incentives to report on positive outcomes when it comes to foreign policy. This will often be sufficient.

Consider, for example, events in Libya in 2011. Some forty-two years earlier, in 1969, Muammar Gaddafi seized power and established a brutal dictatorship. The "mad dog of the Middle East," as President Reagan so aptly called him, was a perpetual thorn in the side of anyone interested in a free society or common human decency.[1] "We have considerable evidence," Reagan said in 1986, "that Qadhafi has been quite outspoken about his participation in urging on and supporting terrorist acts."[2] Despite his history of supporting terrorism, his war crimes, and his calls for countries to adopt "Islamic socialism," it was only in 2011 that the United States government, in conjunction with NATO, finally decided to overthrow him.[3] He was killed in October of that year.

Any reasonable person would applaud the overthrow of Gaddafi. The U.S. government had a responsibility to the Libyan people and their neighbors. Given that Gaddafi had openly threatened the United States, he posed a clear danger. President Obama stated clearly why intervention was necessary:

> The social order in Libya had broken down. ... [Gaddafi had said] "We will kill [the rebels] like rats." Now one option would be to do nothing, and there were some in my administration who said, as tragic as the Libya situation may be, it's not our problem. The way I looked at it was that it would be our problem if, in fact, complete chaos and civil war broke out in Libya. ... At that point, you've got a number of Gulf countries who despise Qaddafi, or are concerned on a humanitarian basis, who are calling for action.[4]

Unfortunately, there were complications. Following Gaddafi's overthrow, the country descended into chaos. A power vacuum emerged, with various militia groups vying for control. More than a decade later, the country remains unstable.[5] Outside of Libya, the whole region has felt the effects of the regime's end.

For instance, Tuareg rebels, who had sought (and received) refuge in Gaddafi's Libya, returned to their home county of Mali following Gaddafi's fall—"heavily armed courtesy of Qaddafi's extensive arsenal."[6] These rebels directly contributed to a coup in Mali in 2012, and the country became a safe haven for terrorist groups.[7]

The same is true for Libya. Strategically located just south of the European continent, Libya has become a safe haven for jihadi terror groups. Not only does the country's instability provide an ideal location for these groups, but also they can easily access weapons. David Lochhead, a senior researcher at the Small Arms Survey, states that the weapons Gaddafi left behind are "still to this day the largest uncontrolled stockpile of ammunition in the world."[8] Libya's neighbors, including Algeria, Egypt, Tunisia, Sudan, Chad, and Niger, have engaged in meetings as recently as 2021 to discuss the problems caused by the country.[9]

Speaking about his time in office, President Obama has gone on record stating that "Libya is a mess." In private, he refers to the country as a "shit show."[10] But Gaddafi's overthrow and the subsequent issues it has presented have been discussed in U.S. media only rarely, if at all. There are few people now, if any, who would call the situation in Libya a blemish on President Obama's legacy.

This is the ideal situation: A major foreign military intervention results in chaos both within the country that was intervened in and in the broader region. Yet, the military operation quickly fades from public memory, with no lasting implications for those who designed and carried out the intervention or for the way that U.S. military primacy is discussed in international affairs.

In the event that failure does not easily fade from the public conscience, you have other options available. One core strategy is to emphasize *current* threats instead of past ones, to remind the public that we can control and shape the future of the world. The past is the past, and we must focus on the present and future. President Obama again serves as an excellent example.

When he announced the conclusion of major combat operations in Iraq in August 2010, President Obama stated, "This milestone should serve as a reminder to all Americans that the future is ours to shape if we move forward with confidence and commitment. It should also serve as a message to the world that the United States of America intends to sustain and strengthen our leadership in this young century."[11] He would later state that it was time to "turn the page" both on the war in Iraq and the debate surrounding it.[12]

President Biden employed a similar strategy when withdrawing troops from Afghanistan. Eleven years to the day after President Obama spoke to the American public about moving on from Iraq, President Biden said, "It's time to look to the future, not the past, to a future that is safer, to a future that is more secure, to a future that honors those who have served."[13] The military and others took the president's charge seriously. Consider that the new Army field operations manual, released in October 2022, mentions Ukraine some sixteen times but fails to mention Afghanistan.[14] This is an important lesson: only look backward on instances of success; otherwise look forward to threats you need to address today and tomorrow.

The good news for you is that because the U.S. government is already involved around the globe, there is always some new conflict that you can leverage to shift public focus from the past.

Consider again the case of Afghanistan. The Biden administration chaotically exited the country in August 2021. Just six months later, in February 2022, the Russian government invaded Ukraine. This created an opportunity to leave the two-decade Afghanistan experience on the ash heap of history as there were renewed calls— both inside and outside America—for the U.S. government to once again reassert and leverage its global dominance in the name of global democracy, peace, and order.

In cases in which you cannot convince the public to move forward and forget the past, you still have other options. For instance, you may elect to cite a lack of funding or a lack of adequate personnel as a reason for the failure. You may also use this as an opportunity to blame your political opponents, who will most likely be the ones refusing to adequately fund your policy. In these cases, the blame shifts easily. Who failed to offer the necessary support for your perfect plan to succeed? *They* did—not you.

A few examples are illustrative. Consider former secretary of state Henry Kissinger. One of the most powerful men in Washington under two different presidential administrations, he approved thousands of bombings in Laos and Cambodia. The bombing of the latter would set the stage for more than a million Cambodians to be murdered by the Khmer Rouge. But, as Kissinger made clear in a master class of blame shifting, funding would have made all the difference in Southeast Asia:

> There is no doubt in my mind that, with anything close to an adequate level of American aid, they would not have collapsed in 1975. And with anything like the support extended to allies in Korea, the Gulf, and the Balkans, they [Cambodia, Laos, and South Vietnam] might have survived until the erosion of communism set it. The end came to Indochina as it does in a Greek tragedy. … The overwhelming Democratic victory in the 1974 congressional elections brought up a group of congressional freshman to Washington who, in the words of *The Almanac of American Politics, 1978*, represented a political realm "in which opposition to the Vietnam War was the most compelling source of motivation."[15]

The Bush administration similarly pointed to funding and political rivals as the causes of issues with the war in Iraq. Once it became apparent that the war was not going to be as quick and decisive as officials had initially promised, the public and President Bush's opponents were quick to call for a drawdown of troops and funding. Bush skillfully used these potentially difficult situations as a means of generating support *and* as a way to shift blame to his political rivals for future problems.

In 2007, for example, Democrats called for funding to be cut and for troops to be withdrawn from Iraq by 2008. In response, President Bush pointed out that, should funding be cut and support withdrawn, *this* would be the source of failure. In the same breath, he tied the refusal to fund his foreign policy with a lack of support for U.S. troops *and* managed to make his adversaries appear lazy—a truly brilliant rhetorical move!

"If Congress fails to pass a bill I can sign by mid-April," President Bush stated, "the Army will be forced to consider cutting back on equipment, equipment repair, and quality of life initiatives for our Guard and Reserve forces. … They need to come off their vacation, get a bill on my desk."[16]

These statements had the desired effect. In May 2007, Congress passed a $120 billion bill to continue operations in Iraq. Though the bill included some eighteen benchmarks "requiring the Iraqi government to make progress on a series of political, economic and security reforms," the president had the discretion to waive those benchmarks.[17] The point about Democrats taking needed supplies from the troops was also roundly heard. George Lisicki, one of the leaders of the national organization Veterans of Foreign Wars, lambasted Senate Majority Leader Harry Reid in the weeks before the bill was passed, stating that withdrawing troops and reducing funding was "reckless and tantamount to waving a white flag of surrender to the enemy. … [This debate] is aimed at the president, but the ultimate target will be the men and women in uniform and their families."[18]

There are other groups you can blame for the failure of foreign policy. For instance, you can blame foreign groups and governments that fail to fall in line with your goals.

When France refused to join the "coalition of the willing" in Iraq and vocally opposed the coalition's operations, government officials were quick to point out that the French government's refusal to lend support hurt the cause of promoting freedom and democracy. As one member of the U.K. Parliament stated, "We need to encourage the French government and President Chirac in particular to seek a role of partnership with the United States, not a position of conflict or tension."[19] Another pointed out that France's refusal to support a United Nations Security Council resolution may have removed more diplomatic options in Iraq. Speaking about France's lack of support for the resolution, one official stated, "Clearly ... had the international community stuck by [the resolution] and sent a strong message of unity to Saddam, that pressure could have borne dividends. *We could have achieved the disarmament that we all want to see and achieved it peacefully.*"[20]

The United States was practically forced to take military action! It's the French government's fault we got into that whole mess. In fact, blaming the French may be a strategy unto itself, with good reason. The Iraq War wasn't the first time the French earned blame for our foreign policy mishaps.

When the United States bombed Libya in response to an attack on our servicemen in 1986, the French stubbornly refused to allow U.S. fighters through their airspace, adding some 1,200 miles to the trip and making the policy more difficult to execute.[21] We also can't forget Vietnam. "*French* Indochina" became a very American problem. In a similar way, the former French colony of Grenada required intervention in 1983 after yet another Marxist dictator attempted to put down roots in Latin America. It was the failure of the French that led the United States to take up the slack and complete the Panama Canal at great cost.

With friends like these, who needs enemies?

Other examples of problematic "allies" abound. Take the most recent U.S. intervention in Libya in 2011. Speaking on the lead-up to military action, President Obama noted that the United States was pressured by its allies: "But what has happened over the last several decades in these circumstances," he said, "is people pushing us to act but then showing an unwillingness to put any skin in the game ... free riders."[22] And once again, after the United States agreed to intervene, we found that our "friends" weren't particularly helpful. "When I got back and ask myself what went wrong [after Gaddafi was overthrown]," President Obama said, "there's room for criticism because I had more faith in the Europeans ... being invested in follow-up. ... [British Prime Minister David Cameron was] distracted by a range of things."[23]

These certainly have not been the only times the United States has been placed in this position by its supposed friends. Unfortunately, we are often the ones, by ourselves, delivering on the "group" project. When things go well, other group members are happy to take the credit. But when things go wrong, they conveniently had nothing to do with the project, or they remind us that they never endorsed the direction of the project in the first place!

Let this be a lesson. Allies are your friends when they fall in line. When they fail to do so, they are friends in name only.

Use this to your advantage to lay blame for any failures that occur. Effective blame shifting will make your life easier domestically while sending a clear message to allies about what happens when they go against you.

There is another important group that can be the target of your blame: the objects of your intervention. Despite being offered a precious gift—freedom and democracy—they are often unwilling, or simply incapable, of receiving it with grace. This is not a new phenomenon. Consider the words of Stokeley W. Morgan, assistant chief of the Division of Latin American Affairs in the State Department. Speaking in 1926, he highlighted how the United States' southern neighbors had been utterly churlish toward their northern neighbor:

> If the United States has received but little gratitude,
> this is only to be expected in a world where grati-
> tude is rarely accorded to the teacher, the doctor,
> or the policeman, and we have been all three. But
> as these young nations grow and develop a greater
> capacity for self-government, and finally take their
> places upon an equal footing with the mature, older
> nations of the world, it may be that in time they will
> come to see the United States with different eyes,
> and to have for her something of the respect and
> affection with which a man regards the instructor
> of his youth and a child looks upon the parent who
> has molded his character.[24]

Things have not improved in the century that has passed since these reflections on the noble work that the United States was attempting in Latin America, either in that region or elsewhere.

For example, in the 1970s, despite the efforts of the United States and others, the Chileans elected Salvador Allende. A Marxist, Allende desired nothing more than to turn the country into a Leninist state. Speaking on the issue, Henry Kissinger placed the blame squarely on the people of Chile for what had happened and what would follow. "I don't see why we need to stand by and watch a country go communist due to the irresponsibility of its own people," he told one of his aides.[25] Kissinger would supervise operations to destabilize Allende, who would be replaced by Augusto Pinochet in a coup d'état. Pinochet's regime would be responsible for gross human rights violations. This isn't surprising, however, given Kissinger's observation and the aforementioned words of Stokeley W. Morgan. The people of Latin America have consistently proved that they are unable to govern themselves and consistently choose poorly in making political decisions.

It's not just those from the United States who recognize this fact. Take the words of Simón Bolívar—"the Liberator" of much of the South American continent. He saw the deficiencies of his fellow South Americans quite clearly in the late nineteenth century:

> [South] America is ungovernable. Those who have served the Revolution have plowed the sea. The only thing that one can do in [South] America is emigrate. These countries shall infallibly pass into the hands of the unleashed multitudes and later almost imperceptibly fall under the power of petty tyrants of all races and colors. ... Perhaps they were beneath the dignity of the Europeans to even conquer them.[26]

It is not only Latin Americans who are incapable of accepting help when they *clearly* require it. To turn once more to President Obama's intervention in Libya in 2011, the former president noted that some of the problems now plaguing the country are the result of the people of Libya. "The degree of tribal division in Libya was greater than our analysts had expected," he said. "And our ability to have any kind of structure there that we could interact with and start providing resources broke down very quickly."[27] He extended this point more broadly, highlighting that many parts of the world—including the Middle East—are largely incapable of constraining these impediments to democracy:

> What has been clear throughout the 20th and 21st centuries is that the progress we make in social order and taming our baser impulses and steadying our fears can be reversed very quickly. Social order starts breaking down. ... Then the default position is tribe—us/them. ... Right now ... you're seeing places that are undergoing severe stress. ... And in those places, the Middle East being exhibit A, the default position ... is to organize tightly in the tribe

and to push back or strike out against those who
are different.

Obama was clearly onto something.

In Iraq, after every attempt was made to export democracy to the
country, the people there were simply unwilling to accept a better
way of life. This fact was recognized by many, including Presidents
Bush and Obama. In 2014, for example, President Obama stated,
"We gave Iraq the chance to have an inclusive democracy."[28] Put
simply, the Iraqis blew it. In fact, a national survey of Iraqis conducted in 2020 found that only 45 percent of respondents thought
that democracy was "the best form of government" for the country.[29]

The same was true for Afghanistan. "We [the U.S. government]
gave the Afghan people a chance to live in a free and democratic
society," President Bush said in a speech in 2005.[30] Sixteen years
later, President Biden echoed similar sentiments following the U.S.
withdrawal:

> We trained and equipped an Afghan military
> force of some 300,000 strong—incredibly well
> equipped—a force larger in size than the militar-
> ies of many of our NATO allies. We gave them
> every tool they could need. We paid their salaries,
> provided for the maintenance of their air force—
> something the Taliban doesn't have. ... We gave
> them every chance to determine their own fu-
> ture. What we could not provide them was the
> will to fight for that future. ... It is wrong to order
> American troops to step up when Afghanistan's
> own armed forces would not.[31]

We gave them all the tools they needed—for *twenty years*—and they
still couldn't figure it out. How can the United States possibly be
responsible?

These examples provide excellent, bipartisan illustrations of how to shift blame to the subjects of your interventions. When shifting blame is done correctly, it is an easy and effective strategy to ensure that their collective guilt insulates you from domestic criticism.

Remember that your policies will not always go according to plan. But it's important to know that these failures are not yours; they are the fault of others. *Always* project perfection, even when things go wrong. Never take blame.

Remember that global order and freedom depend on *you*. Failure is *not* an option.

PART III

How to Use This Guide

11

Running Wars

MEMBERS OF THE national security elite, you shoulder an immense responsibility and burden. *You* are *the* world's savior. *You* are *the* source of order. *You* are *the* fountainhead of freedom and liberty. *You* are *the* god of the state machinery responsible for alleviating the world's ills.

We offer you this manual as a guide for operating an empire and carrying out war for the global good. You must never forget that the U.S. security state is our only hope for world peace. Without your strong hand, we are all doomed to a life of anarchic chaos and violence.

We must stress, however, that this book must not—*cannot*—fall into the wrong hands. You must keep it from those who would use it to sow dissension.

Above all, you must keep this manual from the American public. Failure to do so would be disastrous.

The curtain would be pulled back, and the realities of your operations would be in full view. People would see what is required to run wars and empire under the guise of liberal rhetoric. People would see that you are not all-powerful harbingers of peace but hubristic frauds, hiding behind a curtain while desperately attempting to maintain an outward facade of control over a world that is, in reality, beyond your control.[1] They would see that, for you, spreading and maintaining freedom and liberty is secondary in importance to your maintaining and extending your power and advancing your interests.

People would recognize that running an empire and its related wars requires covert and overt manipulation of the public, neutering the media's ability to serve as an effective check on power, and embracing the creation and dissemination of government propaganda. People would realize that your plans entail the displacement of millions of civilians, which not only creates immediate, significant harms in the here and now, but also plants the seeds for future violent conflict. People would understand that your policies require the maiming and death of innocents who dared to fail to accept the "gift" of liberal democracy and must therefore suffer the consequences.

People would recognize that the operations of liberal empire require a willingness to ignore and violate both international law and ethical precepts of treating other human beings as dignified equals. They would realize that, while trampling on the rights of civilians abroad, they will be relinquishing their own liberties here at home. They would know that their children, their grandchildren, and their great-grandchildren will be saddled with monetary debt on an incomprehensible scale to fund your adventures abroad. Should members of the American public find this guide, they will realize that empire entails the embrace of state capitalism through top-down economic planning and cronyism—the very antithesis of free markets that you purport to value and promote.

If the public reads this manual, they may realize that violence begets violence and that many conflicts today are the result of your past interventions in the affairs of others.

After all, an empire requires being proactively involved throughout the world. The natural result of continuous interventions in complex systems is the emergence of a network of intricate trip wires—hidden triggers that ignite conflict and violence—which can easily be set off, pulling the U.S. government, and hence the U.S. citizens it represents, into a perpetual war for perpetual peace.

To provide an example, consider the January 2024 drone attack on a U.S.-occupied base (Tower 22) located in Jordan, which killed three U.S. soldiers and injured dozens more. The Biden administration blamed Iran for the attack and vowed retaliation (Iranian officials denied responsibility). Why are U.S. troops in Jordan in the first place?

The U.S. government considers Tower 22 to be of strategic value because its far northeastern location borders both Syria and Iraq. This provides the U.S. military access to intervene in both countries in its ongoing conflict with the Islamic State (IS), a group that emerged in 2014 in response to the earlier U.S. occupation of Iraq.[2] At the same time, both the location and general military presence make U.S. personnel vulnerable to violent attacks from an array of enemies. These attacks on U.S. personnel result in calls for retaliatory attacks by the U.S. government; this retaliation, in turn, will likely be met with a violent response in the future, and so the violent spiral continues. Similar U.S.-created trip wires exist around the world.

Recognizing the prevalence and realities of these trip wires— including the recognition that they are often the result of the past actions and decisions of you and your colleagues in the U.S. government—ordinary people might begin to consider the past two and a half decades of their government's actions.

For instance, they might begin to rethink the wisdom of the overly simplistic explanation for the attacks on September 11, 2001— that the perpetrators "hate our freedoms." People might reconsider the costs and consequences of their government's war on terror, which involved military activities in at least eighty-five countries throughout the world and the adoption of police-state practices at home.[3] People might begin to ponder their government's intervention in Libya in 2011 or in the Syrian civil war in 2014 and wonder why their government was involved in these countries while reflecting on the consequences of these actions. People might begin to think about their government's involvement in the ongoing Russo-Ukraine war and in the Israel-Palestine conflict and wonder how we got here and what the endgame might look like. They might consider how their tax dollars are being used to kill innocent men, women, and children absent a clearly articulated plan and reconsider the U.S.'s position as the world's most preeminent arms dealer.

Together, these thoughts might lead people to reconsider the nature of war itself. Rather than thinking about war in terms of simple dichotomies—"good" and "bad"—without nuance and complexity, people might begin to view war in its totality, realizing the brutality and inhumanity of the enterprise of war-making itself.

When they do so, people will begin to question your use of "just war" logic and rhetoric, which typically entails claims and judgments heavily biased in favor of your desired policies and goals.[4] Rather than viewing your wars as necessary acts of heroism, they may come to view the use of violence as a fundamental failure of the system itself—a system that *you* control. This will lead them to consider the morality of the war-making system and all that is required to maintain the apparatus necessary for that system to operate.[5]

We cannot have ordinary, common people thinking for themselves on matters of national security. If they do so, they might realize that there are alternatives to empire and the militaristic imperialism that is empire's outgrowth.[6] They might realize that—while external threats to their persons and their freedoms do indeed exist—empowering you, the national security elite, to protect them also poses a direct threat to those very things. There is a reason, after all, that James Madison warned that "of all the enemies to public liberty, war is perhaps the most to be dreaded, because it comprises and develops the germ of every other" and that "no nation could preserve its freedom in the midst of continual warfare."[7] If people internalize this reality, they will reconsider current war-making institutional arrangements and ask whether elevating militarism as a means of social organization and human interaction, as in the current system, is necessary for and conducive to liberal values and peaceful social cooperation.[8]

The citizenry will also come to realize the limits on what you can accomplish in practice. Just because you, the national security elite, may *want* to achieve particular goals abroad doesn't mean that you *can* achieve these goals.

The public may start to understand the unsurmountable knowledge problems you face: the inability to understand cultures, norms, and how to design institutions in other societies from scratch.[9] People may also begin to see the perverse incentives generated through intervening abroad. They may come to understand how, even *if* you are well intentioned, the self-interest of politicians, foreign government officials, bureaucrats, and special interest groups can (and almost always does) thwart these plans.[10]

Members of the public might come to realize, in other words, that intervening abroad is subject to the same problems, the same inefficiencies, as their least favorite domestic government programs—but even more so! The scale of needed resources, the scope of activities, the number of parties involved, the crossing of international boundaries, and the sheer distance from local circumstances all lay bare the original weaknesses of our policies and amplify them. Problems that were once nationally contained swell to the level of international crisis and catastrophe.

If the American people begin to think for themselves, however, in matters of national security, we run the real risk that they will reject the national security paternalism that they currently embrace, or at least show indifference toward. They may stop demanding from you—as a small child demands from her parents—unquestioned protection from the threat du jour. Upon reflection, people may begin to consider a wide range of other alternatives to militarism as the focal point of international relations: for example, diplomacy, appeasement, trade, and other forms of nongovernment security.[11]

We must not let the public realize that genuine liberal peace, freedom, and prosperity come from the bottom up and not through top-down state planning and control.[12] For people will come to recognize that peace is not a given, single state of affairs resulting from a grand plan, but rather "a constant shaping and reshaping of understandings, situations, and behaviors in a constantly changing lifeworld" between individual human beings.[13] They will realize that peace already exists in all societies because "underneath the layers of violence, each society, without exception, has its peace behaviors, precious resources that can be available to help bring about new and gentler forms of governance locally and on a larger scale in the next century."[14]

We cannot let the citizenry come to appreciate the words and wisdom of those who have opposed war in the past. Otherwise, they may come to believe the words of Jane Addams, who noted after World War I that "it is as practicable to abolish war as it was to abolish chattel slavery."[15] Or they might follow a path similar to that of Major General Smedley Butler, who eventually came to realize that "war is a racket."[16] People may agree with him that we should put our own domestic house in order as opposed to worrying about others. "When our boys were sent off to war they were told it was a 'war to make the world safe for democracy' and a 'war to end all wars,'" he wrote. "Well, eighteen years after, the world has less of a democracy than it had then. Besides, what business is it of ours whether Russia or Germany ... live under democracies or monarchies? Whether they are Fascists or Communists? Our problem is to preserve our own democracy."[17] People might come to think as Albert Einstein did when he argued, in the wake of World War II, that "the only solution is to abolish war and the threat of war. That is the goal toward which we should strive. We must be determined to reject all activities which in any way contradict this goal."[18]

The public must not be allowed to think this way.

Imagine if American citizens entertained the thought that other people might prefer an arrangement different from our own or that you, the national security elite, pose a genuine threat to their most cherished freedoms through your war-making activities. What would happen if ordinary people began to realize that the idea of never-ending liberal global war for the promise of potential liberal peace is inherently contradictory, because employment of the means necessary to carry out such a policy is the very antithesis of liberalism?

People might start to realize that the fear of existential threats is often manufactured.[19] As noted by General Douglas MacArthur in 1957,

> Our government has kept us in a perpetual state of fear—kept us in a continuous stampede of patriotic fervor—with the cry of grave national emergency. Always there has been some terrible evil at home or some monstrous foreign power that was going to gobble us up if we did not blindly rally behind it by furnishing the exorbitant funds demanded. Yet, in retrospect, these disasters seem never to have happened, seem never to have been quite real.[20]

If the public knows what we know, they will come to understand that many of the supposed monsters we have warned them about are not really monsters at all, but convenient cover for growing and maintaining state power under the guise of protecting freedom and liberty.

Nobel laureate economist James Buchanan emphasized that a society cannot be truly free when people are "afraid to be free."[21] His point was that people cannot be free when they are fearful of developing and using their self-governing skills to address difficult challenges. What rubbish!

It is crucial that you continue to perpetuate this fear of total liberty. People must come to see themselves as mere cogs in the state machinery that rules over them and others for their own good because they are incapable of truly governing themselves. This is the only way that this system you operate sustains itself and works!

True freedom can be achieved only under the ever-present boot of the state, ready to stamp down with precision on those who pose a threat to liberty.

At the same time, you must always maintain the rhetoric and image of self-determination, constitutional democracy, and liberty. The legal scholar Michal Glennon has referred to this idea as America's "double government."[22] On the one hand we have the "dignified institutions" that reflect the trappings of democracy and self-determination. These include the traditions, ceremonies, and rituals that define American democracy. On the other hand, we have the "efficient institutions," or what some refer to as the "deep state." This is where the action takes place out of public view. You, the national security elites, are part of the efficient institutions. You get things done when they need to get done.

The key is to never forget how crucial the dignified institutions are to pulling this whole thing off. The best-run wars rely on the Wizard of Oz's motto: "Pay no attention to that man behind the curtain!"

Fortunately for you, it is difficult for a large majority of Americans to envision alternatives to the status quo. This is because you have done an outstanding job convincing them from a young age that empire and militarism are the only legitimate and serious means for responding to global challenges. In the face of these challenges, you have convinced most Americans that you can offer them certainty of safety, freedom, and order.

Bravo! But now is not the time to rest on your laurels.

It is crucial that you continue to perpetuate this myth by emphasizing that alternatives to the status quo are nothing but the fanciful wishes of naive and soft utopians. Failure to do so will create space for the masses to realize that in practice, militarism and empire offer no certain path to beneficial outcomes—in fact, just the opposite. Violence and illiberalism tend to breed violence and illiberalism at home and abroad.[23]

In conclusion, we urge you to remember the words of political commentator H. L. Mencken. Though he was a rabble-rouser and a thorn in the side of the political elite, you can still benefit from his insights. He once wrote, "The most dangerous man to any government is the man who is able to think things out for himself, without regard to the prevailing superstitions and taboos. Almost inevitably he comes to the conclusion that the government he lives under is dishonest, insane and intolerable."[24]

It is for this very reason that this playbook for running wars must be kept confidential and concealed from the common people—to protect them both from others and from themselves. The future of the free world hangs in the balance!

Notes

Chapter 1: A Call to Arms

1. Beckley and Brands 2022.
2. Kagan 2022, n.p.
3. Morgan 1926.
4. Biden 2022b.
5. Biden 2022a, 8.
6. U.S. Department of State 2021.
7. Biden 2022a, 8.
8. Brattberg et al. 2021.
9. Brown 2022.
10. See Haas 2018; Brandt and Wirtschafter 2022.
11. Coyne and Hall 2022.
12. BBC News 2021.
13. Crawford 2019; de la Garza 2022.
14. Biden 2022a, 8.
15. See Charap, Treyger, and Geist 2019.
16. United Nations Office of the High Commissioner for Human Rights 2022.
17. Costs of War 2023.
18. Yayboke 2022.
19. Biden 2022a, 25.
20. U.S. Department of Defense 2022, 5.
21. U.S. Department of Justice 2022.
22. Allen 2022.
23. Atwood 2022.
24. Esfandiari 2023.
25. Biden 2022a, 5.
26. Al Jazeera 2022.

27. Quoted in Zwirko and Kim 2022.
28. Singh 2022.
29. U.S. Department of Homeland Security 2022b.
30. Jefferson 1809.
31. Quoted in Rippon 2020, 101.
32. Chomsky 2015, 63.
33. Quoted in Goldberg 2016.
34. Eisenhower 1954.

Chapter 2: Control the Narrative

1. Quoted in Brown 1976, 10.
2. Joint Chiefs of Staff 2018, iii.
3. Joint Chiefs of Staff 2018, xi.
4. Kennedy 1961.
5. Axelrod 2009.
6. See Winkler 1978.
7. See Turner 1957, 517.
8. See Turner 1957, 517.
9. Quoted in Turner 1957, 515.
10. Roosevelt 1942c.
11. Howell 1997, 795.
12. Quoted in Howell 1997, 796–797.
13. Quoted in Egan 2021.
14. U.S. Department of Homeland Security 2022a.
15. U.S. Department of Homeland Security 2022a.
16. See Hawley 2022.
17. Miyares 2022.
18. Hawley 2022.
19. Quoted in Bond 2022.
20. Quoted in Bailey 1997.
21. Bond 2022.
22. Grabell 2016.
23. See Coyne and Hall 2021, 78–80, 86–95.
24. Quoted in Montgomery 2003.
25. Billingsworth, Butterworth, and Turman 2012, 133.
26. Peter 2020.
27. Peter 2020.
28. Howell 1997, 796–797.
29. U.S. Office of War Information 1942, 1.
30. See Cooper 2001.
31. See Coyne and Hall 2021, 153–156.

32. For a more complete list of films receiving DOD support, see Alford and Secker 2017, 195–204.
  33. Alford and Secker 2017, 195–204.

Chapter 3: Capture the Media

  1. McClellan 2008, 174.
  2. Newman et al. 2022, 12.
  3. Newman et al. 2022, 113.
  4. Newman et al. 2022, 113.
  5. Shearer 2021.
  6. Shearer 2021.
  7. Shearer 2021.
  8. Grandin 2015.
  9. U.S. House of Representatives Committee on Government Reform 2004.
  10. Meet the Press 2002.
  11. Berke and Elder 2001.
  12. Washington Post editors 2003.
  13. Quoted in Bruni 1999.
  14. Quoted in the American Archive of Public Broadcasting 2000.
  15. Bush 2003.
  16. See Coyne and Hall 2021, 54–55.
  17. Coyne and Hall 2021, 54–55.
  18. Washington Post editors 2003.
  19. See Breitman 2015.
  20. See Benedetto 2003.
  21. See Coyne and Hall 2021, 57.
  22. Quoted in Efron and Wright 2002.
  23. Fleischer 2003.
  24. MacAskill 2003.
  25. The Guardian editors 2003.
  26. World Bank 2022.
  27. See Coyne and Hall 2021, 63–64.
  28. Barstow 2008.
  29. Barstow 2008.
  30. Pew Research Center 2003.
  31. Paul and Kim 2004, xxi.
  32. Shafer 2003.
  33. Psaki 2021.
  34. Unknown Author 2021b.
  35. Unknown Author 2021a.
  36. Unknown Author 2021c.

37. Fang 2022.
38. Fang 2022.
39. Fang 2022.
40. White House 2023b.
41. White House 2023b.
42. White House 2023a, emphasis added.
43. White House 2023a.

Chapter 4: Prepare the Sacrifices

1. Tirman 2011.
2. See Johns and Davies 2019.
3. Isikoff 2014.
4. See Sultan and Sediqi 2019.
5. Quoted in Parry and Solomon 2021.
6. Quoted in Byman 2013.
7. Shoker 2018, 1.
8. Shoker 2018, 1.
9. See Dallek 2018.
10. Quoted in Bush 2004, emphasis added.
11. See Watson Institute 2022a.
12. Morrison 2012.
13. See Fetzer et al. 2021, Reifler and Gelpi 2006.
14. Quoted in Committee on Oversight and Government Reform 2008, 48.
15. Quoted in CBS News 2017.
16. Watt 2021.
17. See X 2021c.
18. See X 2021a.
19. See X 2021b.
20. Bush 2005b, emphasis added.
21. Obama 2009.
22. Obama 2009, emphasis added.
23. Cavanaugh 2015, emphasis original.
24. See Kassam 2021 for an example.
25. Philipps 2023.
26. Marine Corps Directorate of Analytics & Performance Optimization 2019.
27. Philipps 2023.
28. Smith 2018.
29. Public Law 113-12, section 2.
30. Kelly, Sergent, and Slack 2019.
31. Absher 2022.
32. Absher 2022.

33. Department of Veterans Affairs 2021, 36.
34. Department of Veterans Affairs 2021, 402.
35. Wong 2019.
36. Krebs and Ralston 2022, 42.
37. U.S. Army 2022.
38. Kershner and Harding 2015.
39. Goldman et al. 2017, 24–25.
40. FWD.us 2022.
41. Population Reference Bureau 2007.
42. Quoted in Johnson 2001.
43. Baldor 2023.
44. See Andrews 2023.
45. Quoted in Myers 2023.
46. Yeung et al. 2017.
47. Ware 2023.
48. Mongilio 2022.
49. Ware 2023.
50. Yeung et al. 2017, 26.
51. U.S. Army Junior ROTC 2023.
52. Baker, Bogel-Burroughs, and Marcus 2022.
53. Quoted in Callander 2000.
54. Quoted in Callander 2000.
55. Baker, Bogel-Burroughs, and Marcus 2022.
56. Quoted in Moses 2010.

Chapter 5: Sacrifice Liberty in the Name of Liberty

1. Quoted in Wilson 2017.
2. Orth 2022.
3. Mueller 2006; Mueller and Stewart 2011.
4. Mueller 2006, 2.
5. Costello and Johnson 2015.
6. For examples, see Eyerman 1998; Abadie 2006; Chenoweth 2010; and Piazza 2013.
7. Posner and Vermeule 2007.
8. Roosevelt 1942b.
9. Roosevelt 1942b.
10. Harry S. Truman Library n.d.
11. Quoted in Coyle 2022.
12. U.S. Government Printing Office 1953.
13. Smith 2021.
14. See Roberts 2020.

15. Roberts 2020.
16. See Satter 2020.
17. For a more detailed overview, see Coyne and Yatsyshina 2021, 119–200.
18. Malhotra 2005.
19. Quoted in Malhotra 2005, 1–2.
20. Malhotra 2005, 3.
21. Public Law 107–56, section 802.
22. American Civil Liberties Union 2023.
23. Public Law 107-56, section 806.
24. American Civil Liberties Union 2023.
25. American Civil Liberties Union 2023.
26. Zavadski 2016.
27. Shane 2015.
28. Quoted in Grim 2012.
29. Shakespeare 1600, Act 3, Scene 5.
30. Pew Research Center 2015.
31. Pew Research Center 2015.
32. Pew Research Center 2015.
33. Gallup 2023a.
34. Gallup 2023a.
35. Associated Press and NORC 2015, 2.
36. Associated Press and NORC 2015, 2.
37. Associated Press and NORC 2015, 3.
38. Associated Press and NORC 2015, 3.
39. Higgs 2006.

Chapter 6: Embrace Top-Down Economic Planning

1. National Park Service 2007, 26.
2. Fagan n.d.
3. U.S. Office of War Information 1942, Fact Sheet No. 12, 1.
4. Wilson 1918.
5. Hitchcock 1918, 545.
6. Hitchcock 1918, 552.
7. Hitchcock 1918, 560.
8. First War Powers Act, 1941.
9. Congressional Research Service 2020, 2.
10. Congressional Research Service 2020, 2.
11. Congressional Research Service 2020, 3.
12. Congressional Research Service 2020, 5.
13. Congressional Research Service 2020, 8.
14. See Siripurapu 2021.

15. Siripurapu 2021.
16. Trump 2020.
17. 3M n.d.
18. 3M n.d.
19. Bergin 2020.
20. Branson et al. 2021.
21. Biden 2021a.
22. Eisenhower 1961.
23. Eisenhower 1961.
24. In economics, public goods are defined by two features: they are nonrivalrous and nonexcludable. Nonrivalry implies that one person's use of a resource does not prevent others from using it. Nonexcludability means that it is impossible (or incredibly costly) to prevent others from using a resource once available. Economists conclude that public goods will be undersupplied absent government intervention.
25. See Coyne 2015.
26. Coyne and Lucas 2016.
27. Smith 2022.
28. Higgs 2007a.
29. Higgs 1987, 241.
30. Priest and Arkin 2011.
31. Flynn 1944; Twight 1975.

Chapter 7: Loosen the Purse Strings

1. Rockoff 2012.
2. Watson Institute 2022b.
3. Savell 2021, 1.
4. Cloud 2016.
5. Cloud 2016.
6. Committee on Finance 1936, 1.
7. Seligman 1918, 7.
8. Forbes 2010.
9. For a discussion of taxes associated with U.S. wars, see Eland 2013.
10. Thorndike 2022.
11. Roosevelt 1942a.
12. Time editors 1942.
13. Higgs 2007b.
14. Disney 1943.
15. Berlin 1942.
16. Marie 2018.
17. Congressional Budget Office 2022, 86.

18. Gallup 2023b.
19. Berlin 1942.
20. See Yang et al. 2017; Linge and Panzer 2022.
21. Andrzejewski 2022.
22. Daggett 2010, 2. The adjustment for 2023 dollars is calculated using the U.S. Bureau of Labor Statistics CPI (Consumer Price Index) Inflation Calculator (https://www.bls.gov/data/inflation_calculator.htm).
23. Cappella 2012, 201–202.
24. Watson Institute 2022b.
25. Kreps 2018, 2.
26. House Bill 4130, Share the Sacrifice Act of 2010.
27. Forbes 2009.
28. Labonte and Levit 2008.
29. Quoted in House Committee on Financial Services 2022.
30. Pelosi 2022.

Chapter 8: Silence Dissent

1. Savage 2023.
2. Chung 2022.
3. Croucher 2018.
4. Byars 2018.
5. Quoted in Bartov 2022.
6. Fifth Congress of the United States 1798.
7. History.com editors 2009.
8. Wilson 1917b, emphasis added.
9. Wilson 1917a.
10. Asp 2022.
11. Greenberg 2010.
12. Coyne, Goodman, and Hall 2019.
13. Apple 1996.
14. Quoted in Nixon Presidential Materials Project 2001.
15. Quoted in Stone and Kuznick 2012, 385.
16. See Myre 2017.
17. See Bradlee 2021.
18. Quoted in Lennard 2022.
19. Quoted in Churchill and Wall 1990, x.
20. See Pilkington 2013.
21. See Hotchner 2011.
22. King 1967.
23. See Gage 2014.
24. Quoted in Meredith 2010, 12.

25. Meredith 2010, 19–20.
26. Quoted in Place 2021.
27. Quoted in Smith 2015.
28. Kelsey 2003.
29. Snapes 2020.
30. BBC News 2010.
31. Morris 2022.
32. Committee on the Judiciary and the Select Subcommittee on the Weaponization of the Federal Government 2023, 6–7.

Chapter 9: Ignore International Law

1. Coyne and Hall 2016.
2. Quoted in Pérez 1979, 484.
3. Freeman 2023.
4. Shepp 2023.
5. McNulty 1999, 268.
6. See Richburg 1993.
7. See Flanagin 2015.
8. Kipling 1899.
9. Ferguson 2004, 198.
10. Prov. 13: 24.
11. Coyne and Hall 2018, 30.
12. Coyne and Hall 2018, 60–63.
13. For a comprehensive overview of the technique, see Rejali 2007, 279.
14. Adams et al. 1902, 65.
15. See Senate Select Committee on Intelligence 2014.
16. Rejali 2007.
17. Rejali 2007, 357–358.
18. Quoted in Schmitt and Marshall 2006.
19. Schmitt and Marshall 2006.
20. Conroy 2005.
21. See Rejali 2007, 292.
22. McKinney 2014.
23. Quoted in Smith 2008.
24. See Rejali 2007, 342–345.
25. See Rejali 2007, 284–285.
26. Ortiz, quoted in Cohn 2012, xi–xii.
27. Monboit 2005.
28. Gottesdiener 2020.
29. Grimmett and Sullivan 2001.
30. Grimmett and Sullivan 2001.

31. Kitfield 1996, 2144.
32. Quoted in Kepner 2001, 488.
33. Gunson 2002.
34. Rohter 2002.
35. United Nations Office for Disarmament Affairs, n.d.
36. Wezeman, Kuimova, and Wezeman 2022, 1.
37. Wezeman, Kuimova, and Wezeman 2022, 2.
38. Wezeman, Kuimova, and Wezeman 2022, 3.
39. For a comprehensive overview of the massacre, see Danner 1994.
40. Scott 2001, 84.
41. Chivers 2016.
42. Chivers 2016.

Chapter 10: Do Not Accept Failure

1. Reagan 1986.
2. Reagan 1986.
3. See Ikiz 2019.
4. Quoted in Goldberg 2016.
5. Council on Foreign Relations 2023.
6. Nossiter 2012.
7. Cole 2020.
8. Quoted in Munshi 2021.
9. Associated Press 2021.
10. Quoted in Goldberg 2016.
11. Obama 2010.
12. Obama 2010.
13. Biden 2021c.
14. U.S. Army 2022.
15. Kissinger 2003, 500.
16. Quoted in CNN 2007.
17. Bresnahan 2007.
18. Quoted in Gurzinski 2007.
19. Quoted in Wintour and Henley 2003.
20. Quoted in Wintour and Henley 2003, emphasis added.
21. Moss 2010, 47.
22. Quoted in Goldberg 2016.
23. Quoted in Goldberg 2016.
24. Morgan 1926, 15.
25. Quoted in Lewis 1975.
26. Quoted in Fox 1971, 59.
27. Quoted in Goldberg 2016.

28. Quoted in Martinez 2014.
29. Center for Insights in Survey Research 2020, 11.
30. Bush 2005a, 992.
31. Biden 2021b.

Chapter 11: Running Wars

1. Whitlock 2021.
2. On the rise of the Islamic State, see Cockburn 2015.
3. Savell 2021, 1; Coyne and Yatsyshina 2021.
4. Fiala 2007, Calhoun 2013.
5. Dobos 2020.
6. Sharp 1990; Schell 2004; Coyne 2022.
7. Madison 1865, 491.
8. These values include individual freedom, human dignity, intellectual humility, voluntary choice and association, freedom of expression, economic freedom, toleration, pluralism, cosmopolitanism, and the commitment to peaceful solutions to interpersonal conflict (see Mises 1996, Kukathas 2003, McCloskey 2019, Boettke 2021).
9. Boettke, Coyne, and Leeson 2008.
10. Coyne 2008, 2013, 2022.
11. Sharp 1990; Mueller 2021; Schell 2004; Coyne and Goodman 2020; Coyne 2022.
12. Coyne 2022, 2023, 2024.
13. Boulding 2000, 1.
14. Boulding 2000, 101.
15. Addams 2006, 285.
16. Butler 2013 [1935]: 25.
17. Butler 2013 [1935]: 50.
18. Einstein 2007, 489.
19. Mueller 2006; Thrall and Cramer 2009.
20. Quoted in Higgs 1994.
21. Buchanan 2005.
22. Glennon 2015.
23. Coyne 2022, 2023.
24. Mencken 1982, 145.

Bibliography

Abadie, Alberto. 2006. "Poverty, Political Freedom, and the Roots of Terrorism." *American Economic Review* 96 (2): 50–56.

Absher, Jim. 2022. "See Your 2023 VA Disability Pay Rates." *Military.com*, November 29, 2022. https://www.military.com/benefits/veterans-health-care/va-disability-pay-rates.html.

Adams, Charles Francis, Carl Schurz, Edwin Burritt Smith, and Herbert Welsh. 1902. *Secretary Root's Record, Marked Severities in Philippine Warfare: Analysis of the Law and Facts Bearing on the Action and Utterances of President Roosevelt and Secretary Root*. Boston: Geo. H. Ellis.

Addams, Jane. 2006. "The Hopes We Inherit." In *Essays and Speeches*, edited by Marilyn Fischer and Judy D. Whipps, 279–286. London: Continuum International Publishing.

Alford, Matthew, and Tom Secker. 2017. *National Security Cinema: The Shocking New Evidence of Government Control in Hollywood*. Scotts Valley, CA: CreateSpace.

Al Jazeera. 2022. "N. Korea Warns of 'All-Out' Nuclear Response to 'U.S. Aggression.'" *Al Jazeera*, November 19, 2022. https://www.aljazeera.com/news/2022/11/19/north-korea-warns-of-all-out-nuclear-response-to-us-provocation.

Allen, Mike. 2022. "Source: Iranian Plot Had $1M Pompeo Bounty." *Axios*, August 10, 2022. https://www.axios.com/2022/08/10/iran-assassination-pompeo-bolton.

American Archive of Public Broadcasting. 2000. *The NewsHour with Jim Lehrer*, February 16, 2000. https://americanarchive.org/catalog/cpb-aacip-507-b56d21s522.

American Civil Liberties Union. 2023. "How the USA PATRIOT Act Redefines 'Domestic Terrorism.'" https://www.aclu.org/other/how-usa-patriot-act-redefines-domestic-terrorism.

Andrews, Lena S. 2023. *Valiant Women: The Extraordinary American Servicewomen Who Helped Win World War II*. New York: HarperCollins.

Andrzejewski, Adam. 2022. "NSA Wasted $3.6M on Unused Parking Garage." *RealClear Policy*, May 20, 2022. https://www.realclearpolicy.com/articles/2022/05/20/nsa_wasted_36m_on_unused_parking_garage_832067.html.

Apple, R. W., Jr. 1996. "25 Years Later; Lessons from the Pentagon Papers." *New York Times*, June 23, 1996. https://www.nytimes.com/1996/06/23/weekinreview/25-years-later-lessons-from-the-pentagon-papers.html.

Asp, David. 2022. "Espionage Act of 1917." *The First Amendment Encyclopedia*. https://www.mtsu.edu/first-amendment/article/1045/espionage-act-of-1917.

Associated Press. 2021. "Libya's Neighbors Meet, Urge Foreign Fighters to Leave." *Associated Press*, August 31, 2021. https://apnews.com/article/middle-east-africa-libya-84e9187a0039c94a2226f2de0a302859.

Associated Press and NORC. 2015. "Americans Evaluate the Balance Between Security and Civil Liberties." https://apnorc.org/wp-content/uploads/2020/02/2015-12-Security-and-Civil-Liberties_FINAL.pdf.

Atwood, Kylie. 2022. "Iran Is Preparing to Send Additional Weapons Including Ballistic Missiles to Russia to Use in Ukraine, Western Officials Say." *CNN Politics*, November 1, 2022. https://www.cnn.com/2022/11/01/politics/iran-missiles-russia/index.html.

Axelrod, Alan. 2009. *Selling the Great War: The Making of American Propaganda*. New York: Palgrave Macmillan.

Bailey, Ronald. 1997. "The Origin of the Specious: Why Do Conservatives Doubt Darwin?" *Reason*, July 1997. https://reason.com/1997/07/01/origin-of-the-specious/.

Baker, Mike, Nicholas Bogel-Burroughs, and Ilana Marcus. 2022. "Thousands of Teens Are Being Pushed into Military's Junior R.O.T.C." *New*

York Times, December 11, 2022. https://www.nytimes.com/2022/12/11/us/jrotc-schools-mandatory-automatic-enrollment.html.

Baldor, Lolita C. 2023. "Join the Military, Become a U.S. Citizen: Uncle Sam Wants You and Vous and Tu." *Associated Press*, June 11, 2023. https://apnews.com/article/army-air-force-recruiting-shortfall-immigrants-citizenship-2cd690352210606945010d1800c5bdbe.

Barstow, David. 2008. "Behind TV Analysts, Pentagon's Hidden Hand." *New York Times*, April 20, 2008. https://www.nytimes.com/2008/04/20/us/20generals.html.

Bartov, Shira Li. 2022. "Teacher Caught Yelling at Student over Pledge of Allegiance in Viral Video." *Newsweek*, September 23, 2022. https://www.newsweek.com/teacher-caught-yelling-student-pledge-allegiance-viral-tiktok-florida-1745887.

BBC News. 2021. "Report: China Emissions Exceed All Developed Nations Combined." *BBC News*, May 7, 2021. https://www.bbc.com/news/world-asia-57018837.

———. 2010. "PayPal Says It Stopped WikiLeaks Payments on US Letter." *BBC News*, December 8, 2010. https://www.bbc.com/news/business-11945875.

Beckley, Michael, and Hal Brands. 2022. "The Return of Pax Americana?" *Foreign Affairs*, March 14, 2022. https://www.foreignaffairs.com/articles/russia-fsu/2022-03-14/return-pax-americana.

Benedetto, Richard. 2003. "Poll: Most Back War, but Want U.N. Support." *USA Today*, March 16, 2003. https://web.archive.org/web/20120326155359/http://www.usatoday.com/news/world/iraq/2003-03-16-poll-iraq_x.htm.

Bergin, Tom. 2020. "The U.S. Has Spent Billions Stockpiling Ventilators, but Many Won't Save Critically Ill COVID-19 Patients." *Reuters*, December 2, 2020. https://www.reuters.com/article/us-health-coronavirus-ventilators-insigh/the-u-s-has-spent-billions-stockpiling-ventilators-but-many-wont-save-critically-ill-covid-19-patients-idUSKBN28C1N6.

Berke, Richard L., and Janet Elder. 2001. "Poll Finds Support for War and Fear on the Economy." *New York Times*, September 25, 2001. https://www.nytimes.com/2001/09/25/us/a-nation-challenged-the-poll-poll-finds-support-for-war-and-fear-on-economy.html.

Berlin, Irving. 1942. "I Paid My Income Tax Today." National Archives. http://s.wsj.net/public/resources/documents/WSJ_Paid_Income_Tax041408.pdf.

Biden, Joseph. 2022a. "National Security Strategy." October 12, 2022. https://www.whitehouse.gov/wp-content/uploads/2022/10/Biden-Harris-Administrations-National-Security-Strategy-10.2022.pdf.

———. 2022b. "Remarks by the President on Standing Up for Democracy." November 2, 2022. https://www.whitehouse.gov/briefing-room/speeches-remarks/2022/11/03/remarks-by-president-biden-on-standing-up-for-democracy/.

———. 2021a. "Executive Order on a Sustainable Public Health Supply Chain." January 21, 2021. https://www.whitehouse.gov/briefing-room/presidential-actions/2021/01/21/executive-order-a-sustainable-public-health-supply-chain/.

———. 2021b. "Remarks by President Biden on Afghanistan." August 16, 2021. https://www.whitehouse.gov/briefing-room/speeches-remarks/2021/08/16/remarks-by-president-biden-on-afghanistan/.

———. 2021c. "Remarks by President Biden on the End of the War in Afghanistan." August 31, 2021. https://www.whitehouse.gov/briefing-room/speeches-remarks/2021/08/31/remarks-by-president-biden-on-the-end-of-the-war-in-afghanistan/.

Billingsworth, Andrew C., Michael L. Butterworth, and Paul D. Turman. 2012. *Communication and Sport: Surveying the Field.* New York: Sage.

Boettke, Peter J. 2021. *The Struggle for a Better World.* Arlington: Mercatus Center at George Mason University.

Boettke, Peter J., Christopher J. Coyne, and Peter T. Leeson. 2008. "Institutional Stickiness and the New Development Economics," *American Journal of Economics and Sociology* 67(2): 331–358.

Bond, Shannon. 2022. "She Joined DHS to Fight Disinformation. She Was Halted by … Disinformation." *NPR*, May 21, 2022. https://www.npr.org/2022/05/21/1100438703/dhs-disinformation-board-nina-jankowicz.

Boulding, Elise. 2000. *Cultures of Peace: The Hidden Side of History.* Syracuse, NY: Syracuse University Press.

Bradlee, Ben, Jr. 2021. "The Deceit and Conflict Behind the Leak of the Pentagon Papers." *New Yorker*, April 8, 2021. https://www.newyorker.com/news/american-chronicles/the-deceit-and-conflict-behind-the-leak-of-the-pentagon-papers.

Brandt, Jessica, and Valerie Wirtschafter. 2022. "China Uses Search Engines to Spread Propaganda." *Tech Stream*, July 6, 2022. https://www.brookings. edu/techstream/how-china-uses-search-engines-to-spread-propaganda/.

Branson, Rich, Jeffrey R. Dichter, Henry Feldman, Asha Devereaux, David Dries, Joshua Benditt, Tanzib Hossain, Marya Ghazipura, Mary King, Marie Baldisseri, Michael D. Christian, Guillermo Dominguez-Cherit, Kiersten Henry, Anne Marie O. Martland, Meredith Huffines, Doug Ornoff, Jason Persoff, Dario Rodriguez Jr., Ryan C. Maves, Niranjan "Tex" Kissoon, and Lewis Rubinson. 2021. "The U.S. Strategic National Stockpile Ventilators in Coronavirus Disease 2019." *Chest* 159 (2): 634–652.

Brattberg, Erik, Phillippe Le Corre, Paul Stronski, and Thomas de Waal. 2021. "China's Influence in Southeastern, Central, and Eastern Europe." Carnegie Endowment for International Peace. https://carnegieendowment.org/ files/202110-Brattberg_et_al_EuropeChina_final.pdf.

Breitman, Kendall. 2015. "Poll: Half of Republicans Still Believe WMDs Found in Iraq." *Politico*, January 7, 2015. https://www.politico.com/ story/2015/01/poll-republicans-wmds-iraq-114016.

Bresnahan, John. 2007. "Congress Passes $120 Billion Iraq Bill." *Politico*, May 24, 2007. https://www.politico.com/story/2007/05/congress-passes-120-billion-iraq-bill-004178.

Brown, Anthony Cave. 1976. *Bodyguard of Lies: The Classic History of the War of Deception That Kept D-Day Secret from Hitler and Sealed the Allied Victory.* New York: Harper Perennial.

Brown, David. 2022. "Why China Could Win the New Global Arms Race." *BBC News*, July 28, 2022. https://www.bbc.com/news/ world-asia-china-59600475.

Bruni, Frank. 1999. "Bush Has Tough Words and Rough Enunciation for Iraqi Chief." *New York Times*, December 4, 1999. https://www.nytimes. com/1999/12/04/us/bush-has-tough-words-and-rough-enunciation-for-iraqi-chief.html.

Buchanan, James M. 2005. "Afraid to Be Free: Dependency as Desideratum," *Public Choice* 124 (1):19–31.

Bush, George W. 2005a. "Remarks at the President's Dinner, June 14." In *Public Papers of the Presidents of the United States: George W. Bush,* Book I, 988–992.Washington, DC: U.S. Government Publication Office.

———. 2005b. "Speech at Fort Bragg, North Carolina." June 28, 2005. https://www.theguardian.com/world/2005/jun/29/iraq.usa.

———. 2004. "Transcript from Bush Speech on American Strategy in Iraq." *New York Times*, May 24, 2004. https://www.nytimes.com/2004/05/24/politics/transcript-from-bush-speech-on-american-strategy-in-iraq.html.

———. 2003. "President's Radio Address." February 8, 2003. https://georgewbush-whitehouse.archives.gov/news/releases/2003/02/20030208.html.

Butler, Smedley. 2013 [1935]. *War Is a Racket*. New York: Skyhorse Publishing.

Byars, Mitchell. 2018. "Colorado Teacher Pleads Guilty to Child Abuse After Forcing Student to Stand for Pledge of Allegiance." *Cañon City Daily Record*, August 30, 2018. https://www.canoncitydailyrecord.com/2018/08/30/colorado-teacher-pleads-guilty-to-child-abuse-after-forcing-student-to-stand-for-pledge-of-allegiance/.

Byman, Daniel L. 2013. "Why Drones Work: The Case for Washington's Weapon of Choice." *Brookings*, June 17, 2013. https://www.brookings.edu/articles/why-drones-work-the-case-for-washingtons-weapon-of-choice/.

Calhoun, Laurie. 2013. *War and Delusion: A Critical Examination*. New York: Palgrave Macmillan.

Callander, Bruce D. 2000. "The Surge in Junior ROTC." *Air and Space Forces Magazine*, April. https://www.airandspaceforces.com/article/0400rotc/.

Cappella, Rosella. 2012. "The Political Economy of War Finance." Dissertation, University of Pennsylvania. https://repository.upenn.edu/cgi/viewcontent.cgi?article=2985&context=edissertations.

Cavanaugh, Matt L. 2015. "A Soldier Can Never Die in Vain." *War on the Rocks*, October 12, 2015. https://warontherocks.com/2015/10/a-soldier-can-never-die-in-vain/.

CBS News. 2017. "Pat Tillman's Widow to Trump: Don't Politicize Husband's Service." *CBS News*, September 26, 2017. https://www.cbsnews.com/sanfrancisco/news/pat-tillman-widow-to-president-trump-dont-politicize-husband-service/.

Center for Insights in Survey Research. 2020. *Nationwide Public Opinion Poll of Iraqi Citizens*. International Republican Institute. https://www.iri.org/wp-content/uploads/2020/12/irq-20-ns-01-pt-public.pdf.

Charap, Samuel, Elina Treyger, and Edward Geist. 2019. *Understanding Russia's Intervention in Syria*. RAND Corporation. https://www.rand.org/pubs/research_reports/RR3180.html.

Chenoweth, E. 2010. "Democratic Pieces: Democratization and the Origins of Terrorism." In *Coping with Terrorism: Origins, Escalation, Counterstrategies, and Responses*, edited by R. Reuveny and W. R. Thompson, 97–123. Albany, NY: State University of New York Press.

Chivers, C. J. 2016. "How Many Guns Did the U.S. Lose Track of in Iraq and Afghanistan? Hundreds of Thousands." *New York Times Magazine*, August 24, 2016. https://www.nytimes.com/2016/08/23/magazine/how-many-guns-did-the-us-lose-track-of-in-iraq-and-afghanistan-hundreds-of-thousands.html.

Chomsky, Noam. 2015. *On Power and Ideology: The Managua Lectures*. Chicago: Haymarket Books.

Chung, Christine. 2022. "Texas Student Who Protested Pledge of Allegiance Gets $90,000 in Settlement." *New York Times*, March 31, 2022. https://www.nytimes.com/2022/03/31/us/texas-pledge-of-allegiance-lawsuit.html.

Churchill, Ward, and Jim Vander Wall. 1990. *The COINTELPRO Papers: Documents from the FBI's Secret Wars Against Domestic Dissent*. Boston: South End Press.

Cloud, David S. 2016. "How Much Do Allies Pay for U.S. Troops? A Lot More than Donald Trump Says." *Los Angeles Times*, October 1, 2016. https://www.latimes.com/nation/la-na-trump-allies-20160930-snap-story.html.

CNN. 2007. "Bush: Clock Ticking on Funding for War Troops." *CNN*, April 3. https://www.cnn.com/2007/POLITICS/04/03/senate.funds/index.html.

Cockburn, Patrick. 2015. *The Rise of Islamic State: ISIS and the New Sunni Revolution*. London: Verso.

Cohn, Marjorie (ed.). 2012. *The United States and Torture*. New York: New York University Press.

Cole, Emily. 2020. "Five Things to Know About Mali's Coup." *United States Institute of Peace*, August 27, 2020. https://www.usip.org/publications/2020/08/five-things-know-about-malis-coup.

Committee on Finance. 1936. *War Revenue Act: Report from the Committee on Finance.* 74th Congress, Second Session. https://www.finance.senate.gov/imo/media/doc/74PrtWarRevenue.pdf.

Committee on Oversight and Government Reform. 2008. *Misleading Information from the Battlefield: The Tillman and Lynch Episodes.* https://www.govinfo.gov/content/pkg/CRPT-110hrpt858/pdf/CRPT-110hrpt858.pdf.

Committee on the Judiciary and the Select Subcommittee on the Weaponization of the Federal Government. 2023. *The Weaponization of the Federal Trade Commission: An Agency's Overreach to Harass Elon Musk's Twitter.* March 7, 2023. https://judiciary.house.gov/sites/evo-subsites/republicans-judiciary.house.gov/files/evo-media-document/Weaponization_Select_Subcommittee_Report_on_FTC_Harrassment_of_Twitter_3.7.2023.pdf.

Congressional Budget Office. 2022. *The Budget and Economic Outlook: 2022 to 2032.* https://www.cbo.gov/system/files/2022-05/57950-Outlook.pdf.

Congressional Research Service. 2020. *The Defense Production Act of 1950: History, Authorities, and Considerations for Congress.* https://sgp.fas.org/crs/natsec/R43767.pdf.

Conroy, John. 2005. "Tools of Torture." *Chicago Reader*, February 3, 2005. https://chicagoreader.com/news-politics/tools-of-torture/.

Cooper, Marc. 2001. "Lights! Camera! Attack! Hollywood Enlists." *The Nation,* November 21, 2001. https://www.thenation.com/article/archive/lights-cameras-attack-hollywood-enlists/.

Costello, Tom, and Alex Johnson. 2015. "TSA Chief Out After Agents Fail 95 Percent of Airport Breach Tests." *NBC News*, June 1, 2015. https://www.nbcnews.com/news/us-news/investigation-breaches-us-airports-allowed-weapons-through-n367851.

Costs of War. 2023. "Human Cost of Post-9/11 Wars: Direct War Deaths in Major War Zones, Afghanistan & Pakistan (Oct. 2001–Aug. 2021); Iraq (March 2003–March 2023); Syria (Sept. 2014–March 2023); Yemen (Oct. 2002–Aug. 2021) and Other Post-9/11 War Zones." https://watson.brown.edu/costsofwar/figures/2021/WarDeathToll.

Council on Foreign Relations. 2023. "Civil Conflict in in Libya." Updated September 19, 2023. https://www.cfr.org/global-conflict-tracker/conflict/civil-war-libya.

Coyle, Jake. 2022. "Marsha Hunt, '40s Star and Blacklist Victim, Dies at 104." *AP News*, September 10, 2022. https://apnews.com/article/entertainment-movies-california-c1945afdd00c51db0202a8f68cc4fe3c.

Coyne, Christopher J. 2024. "Peacemaking: Top-Down vs. Bottom-Up." *Markets & Society*, forthcoming.

———. 2023. "The Folly of Empire and the Science of Peace." *Journal of Private Enterprise* 38 (2): 1–15.

———. 2022. *In Search of Monsters to Destroy: The Follies of American Empire and Paths to Peace*. Oakland, CA: Independent Institute.

———. 2015. "Lobotomizing the Defense Brain." *Review of Austrian Economics* 28 (4): 371–396.

———. 2013. *Doing Bad by Doing Good: Why Humanitarian Action Fails*. Stanford, CA: Stanford University Press.

———. 2008. *After War: The Political Economy of Exporting Democracy*. Stanford, CA: Stanford University Press.

Coyne, Christopher J., and Abigail R. Hall. 2022. "Dr. Mengele, USA Style: Lessons from Human Rights Abuses in Post–World War II America." *Cosmos+Taxis* 10 (9+10): 113–126.

———. 2021. *Manufacturing Militarism: U.S. Government Propaganda in the War on Terror*. Stanford, CA: Stanford University Press.

———. 2018. *Tyranny Comes Home: The Domestic Fate of U.S. Militarism*. Stanford, CA: Stanford University Press.

———. 2016. "Empire State of Mind: The Illiberal Foundation of Liberal Hegemony." *Independent Review* 21 (2): 237–250.

Coyne, Christopher J., and David S. Lucas. 2016. "Economists Have No Defense: A Critical Review of National Defense in Economics Textbooks." *Journal of Private Enterprise* 31 (4): 65–83.

Coyne, Christopher J., and Nathan P. Goodman. 2020. "Polycentric Defense." *Independent Review* 25 (2): 279–292.

Coyne, Christopher J., Nathan Goodman, and Abigail R. Hall. 2019. "Sounding the Alarm: The Political Economy of Whistleblowing in the US Security State." *Peace Economics, Peace Science, and Public Policy* 25 (1): 1–11.

Coyne, Christopher J., and Yuliya Yatsyshina. 2021. "Police State, U.S.A." *Independent Review* 26 (2): 189–204.

Crawford, Neta C. 2019. "Pentagon Fuel Use, Climate Change, and the Costs of War." *Costs of War*, November 13, 2019. https://watson.brown.edu/

costsofwar/files/cow/imce/papers/Pentagon%20Fuel%20Use%2C%20
Climate%20Change%20and%20the%20Costs%20of%20War%20
Revised%20November%202019%20Crawford.pdf.

Croucher, Shane. 2018. "Texas Attorney General Takes on Black Teen
Expelled for Sitting During Pledge of Allegiance." *Newsweek*, September
17, 2018. https://www.newsweek.com/texas-black-student-expelled-
pledge-allegiance-attorney-general-1140412.

Daggett, Stephen. 2010. *Costs of Major U.S. Wars*. Congressional Research
Service, June 29, 2010. https://apps.dtic.mil/sti/pdfs/ADA524288.pdf.

Dallek, Matthew. 2018. "How the Army's Cover-Up Made the My Lai Massacre Even Worse." *History.com*, August 30, 2018. https://www.history.
com/news/my-lai-massacre-1968-army-cover-up.

Danner, Mark. 1994. *The Massacre at El Mozote*. New York: Vintage.

de la Garza, Alejandro. 2022. "To Take Climate Change Seriously, the
U.S. Military Needs to Shrink." *Time*, February 17, 2022. https://time.
com/6148778/us-military-climate-change/.

Disney, Walt. 1943. *The Spirit of '43*. https://www.youtube.com/
watch?v=i4lFj2wp8do.

Dobos, Ned. 2020. *Ethics, Security, and the War-Machine: The True Cost of the
Military*. New York: Oxford University Press.

Efron, Sonni, and Robin Wright. 2002. "Blair Urges Action on Iraq." *Los
Angeles Times*, September 8, 2002. https://www.latimes.com/archives/la-
xpm-2002-sep-08-fg-usiraq8-story.html.

Egan, Lauren. 2021. "'They're Killing People': Biden Blames Facebook, Other
Social Media for Allowing Covid Misinformation." *NBC News*, July 16,
2021. https://www.nbcnews.com/politics/white-house/they-re-killing-
people-biden-blames-facebook-other-social-media-n1274232.

Einstein, Albert. 2007. *Einstein on Politics: His Private Thoughts and Public
Stands on Nationalism, Zionism, War, Peace, and the Bomb*. Princeton, NJ:
Princeton University Press.

Eisenhower, Dwight D. 1961. Farewell Address. January 17, 1961. National Archives. https://www.archives.gov/milestone-documents/
president-dwight-d-eisenhowers-farewell-address.

———. 1954. Statement at Press Conference on April 7, 1954. Office of the Historian. https://history.state.gov/historicaldocuments/
frus1952-54v13p1/d716.

Eland, Ivan. 2013. "Warfare State to Welfare State: Conflict Causes Government to Expand at Home." *Independent Review* 18 (2): 189–218.

Esfandiari, Haleh. 2023. "Hamas and Israel: Iran's Role." Wilson Center, October 10, 2023. https://www.wilsoncenter.org/article/hamas-and-israel-irans-role#:~:text=The%20Iranian%20role%20in%20the,by%20training%20is%20well%20known.

Eyerman, J. (1998). "Terrorism and Democratic States: Soft Targets or Accessible Systems." *International Interactions* 24 (2): 151–170.

Fagan, Shemia. n.d. "Rationing: A Necessary but Hated Sacrifice." Oregon Secretary of State. https://sos.oregon.gov/archives/exhibits/ww2/Pages/services-rationing.aspx.

Fang, Lee. 2022. "Twitter Aided the Pentagon in Its Covert Online Propaganda Campaign." *The Intercept*, December 20, 2022. https://theintercept.com/2022/12/20/twitter-dod-us-military-accounts/.

Ferguson, Niall. 2004. *Colossus: The Rise and Fall of the American Empire.* New York: Penguin Books.

Fetzer, Thiemo, Pedro C. L. Souza, Oliver Vanden Eynde, and Austin L. Wright. 2021. "Losing on the Home Front? Battlefield Casualties, Media, and Public Support for Foreign Interventions." Becker Friedman Institute Working Paper No. 2021-52. https://bfi.uchicago.edu/wp-content/uploads/2021/04/BFI_WP_2021-52.pdf.

Fiala, Andrew. 2007. *The Just War Myth: The Moral Illusions of War.* Lanham, MD: Rowman and Littlefield.

Fifth Congress of the United States. 1798. "The Alien and Sedition Acts." https://www.archives.gov/milestone-documents/alien-and-sedition-acts#transcript.

First War Powers Act. 1941. 50a U.S.C. §§ 601–622 (1946). https://tile.loc.gov/storage-services/service/ll/uscode/uscode1946-00405/uscode1946-004050a009/uscode1946-004050a009.pdf.

Flanagin, Jake. 2015. "Bill Clinton Failed to Stop Genocide in Rwanda, but His Foundation Is Quietly Making Amends." *Quartz*, April 16, 2015. https://qz.com/384228/the-clinton-foundation-is-atoning-for-bills-failure-on-rwanda.

Fleischer, Ari. 2003. "Excerpts from the Press Briefing by Ari Fleischer." March 18, 2003. https://georgewbush-whitehouse.archives.gov/infocus/iraq/news/20030318-5.html.

Flynn, John T. 1944. *As We Go Marching*. New York: Free Life Editions.

Forbes. 2010. "A Short History of Taxes." *Forbes*, April 14, 2010. https://www. forbes.com/2010/04/14/tax-history-law-personal-finance-tax-law-changes. html?sh=3164fba41cf8.

———. 2009. "The Cost of War." *Forbes*, November 26, 2009. https://www. forbes.com/2009/11/25/shared-sacrifice-war-taxes-opinions-columnists-bruce-bartlett.html?sh=214d721d33df.

Fox, Hugh. 1971. "The Bitter Legacy of Bolívar." *Southwest Review* 56 (1): 55–64.

Freeman, Will. 2023. "Why the Situation in Cuba Is Deteriorating." Council on Foreign Relations, April 25, 2023. https://www.cfr.org/in-brief/ why-situation-cuba-deteriorating.

FWD.us. 2022. "5 Things to Know About Immigrants in the Military." September 14, 2022. https://www.fwd.us/news/immigrants-in-the-military/.

Gage, Beverly. 2014. "What an Uncensored Letter to M.L.K. Reveals." *New York Times Magazine*, November 11, 2014. https://www.nytimes. com/2014/11/16/magazine/what-an-uncensored-letter-to-mlk-reveals. html.

Gallup. 2023a. "Civil Liberties." https://news.gallup.com/poll/5263/civil-liberties.aspx.

———. 2023b. "Taxes." https://news.gallup.com/poll/1714/taxes.aspx.

Glennon, Michael J. 2015. *National Security and Double Government*. New York: Oxford University Press.

Goldberg, Jeffrey. 2016. "The Obama Doctrine: The President Explains His Hardest Decisions About America's Role in the World." *The Atlantic*, April 2016. https://www.theatlantic.com/magazine/archive/2016/04/ the-obama-doctrine/471525/.

Goldman, Charles A., Jonathan Schweig, Maya Beunaventura, and Cameron Wright. 2017. *Geographic and Demographic Representativeness of Junior Reserve Officer Training Corps*. RAND Corporation. https://www.rand. org/content/dam/rand/pubs/research_reports/RR1700/RR1712/RAND_ RR1712.pdf.

Gottesdiener, Laura. 2020. "The Children of Fallujah: The Medical Mystery at the Heart of the Iraq War." *The Nation*, November 9, 2020. https://www. thenation.com/article/world/fallujah-iraq-birth-defects/.

Grabell, Michael. 2016. "The TSA Releases Data on Air Marshal Misconduct, 7 Years After We Asked." *ProPublica*, February 24, 2016. https://www. propublica.org/article/tsa-releases-data-on-air-marshal-misconduct-7-years-after-we-asked.

Grandin, Greg. 2015. "Henry Kissinger's Genocidal Legacy: Vietnam, Cambodia and the Birth of American Militarism." *Salon*, November 10, 2015. https://www.salon.com/2015/11/10/ henry_kissingers_genocidal_legacy_partner/.

Greenberg, David. 2010. "The Hidden History of the Espionage Act." *Slate*, December 27, 2010. https://slate.com/news-and-politics/2010/12/the-real-purpose-of-the-espionage-act.html.

Grim, Ryan. 2012. "Robert Gibbs Says Anwar al-Awlaki's Son, Killed by Drone Strike, Needs 'Far More Responsible Father.'" *HuffPost*, October 24, 2012. https://www.huffpost.com/entry/robert-gibbs-anwar-al-awlaki_n_2012438.

Grimmett, Richard F., and Mark P. Sullivan. 2001. *U.S. Army School of the Americas: Background and Congressional Concerns*. Congressional Research Service, April 16, 2001. https://apps.dtic.mil/sti/pdfs/ADA527172.pdf.

Gunson, Phil. 2002. "Hugo Banzer." *The Guardian*, May 5, 2002. https:// www.theguardian.com/news/2002/may/06/guardianobituaries.bolivia.

Gurzinski, John. 2007. "Bush, Democrats Play Blame Game Over War Funding." *Las Vegas Review-Journal*, April 4, 2007. https://www.reviewjournal. com/news/bush-democrats-play-blame-game-over-war-funding/.

Haas, Benjamin. 2018. "China Bans Winnie the Pooh Film After Comparisons to President Xi." *The Guardian*, August 6, 2018. https://www.theguardian. com/world/2018/aug/07/china-bans-winnie-the-pooh-film-to-stop-comparisons-to-president-xi.

Harry S. Truman Library. n.d. "Japanese-American Internment." https://www.trumanlibrary.gov/education/presidential-inquiries/ japanese-american-internment.

Hawley, Josh. 2022. "Hawley, Grassley Demand Answers on New Whistleblower Documents Exposing DHS Disinformation Board Efforts to Monitor Americans' Speech." June 8, 2022. https://www.hawley.senate. gov/hawley-grassley-demand-answers-new-whistleblower-documents-exposing-dhs-disinformation-board.

Higgs, Robert. 2007a. "Military-Economic Fascism: How Business Corrupts Government, and Vice Versa." *Independent Review* 12 (2): 299–316.

———. 2007b. "Wartime Origins of Modern Income-Tax Withholding." *Independent Institute*, December 24, 2007. https://www.independent.org/news/article.asp?id=2092.

———. 2006. "Fear: The Foundation of Every Government's Power." *Independent Review* 10 (3): 447–466.

———. 1994. "The Cold War Economy: Opportunity Costs, Ideology, and the Politics of Crisis." *Independent Institute*, July 1994. https://www.independent.org/publications/article.asp?id=1297.

———. 1987. *Crisis and Leviathan: Critical Episodes in the Growth of American Government*. New York: Oxford University Press.

History.com editors. 2009. "Alien and Sedition Acts." *History.com*. https://www.history.com/topics/earlyus/alien-and-sedition-acts.

Hitchcock, Curtice N. 1918. "The War Industries Board: Its Development, Organization, and Functions." *Journal of Political Economy* 26 (6): 545–566.

Hotchner, A. E. 2011. "Hemingway, Hounded by the Feds." *New York Times*, July 1, 2011. https://www.nytimes.com/2011/07/02/opinion/02hotchner.html.

House Bill 4130, Share the Sacrifice Act of 2010. https://www.congress.gov/bill/111th-congress/house-bill/4130/text?r=4&s=1.

House Committee on Financial Services. 2022. "The Inflation Equation: Corporate Profiteering, Supply Chain Bottlenecks, and COVID-19." U.S. Government Publishing Office, March 8, 2022. https://www.congress.gov/event/117th-congress/house-event/114484/text.

Howell, Thomas. 1997. "The Writer's War Board: U.S. Domestic Propaganda in World War II." *The Historian* 59 (4): 795–813.

Ikiz, Ahmet Salih. 2019. "Green Book and Islamic Socialism." *Eurasian Journal of Social Sciences* 7 (3): 23–29.

Isikoff, Michael. 2014. "Yemenis: Drone Strike 'Turned Wedding into Funeral.'" *NBC News*, January 7, 2014. https://www.nbcnews.com/news/investigations/yemenis-drone-strike-turned-wedding-funeral-n5781.

Jefferson, Thomas. 1809. "Thomas Jefferson to James Madison, 27 April 1809." https://founders.archives.gov/documents/Jefferson/03-01-02-0140.

Johns, Robert, and Graeme A. M. Davies. 2019. "Civilian Casualties and Public Support for Military Action: Experimental Evidence." *Journal of Conflict Resolution* 63 (1): 251–281.

Johnson, Greg. 2001. "Enlisting Spanish to Recruit the Troops." *Los Angeles Times*, March 14, 2001. https://www.latimes.com/archives/la-xpm-2001-mar-14-fi-37501-story.html.

Joint Chiefs of Staff. 2018. *Joint Concept for Operation in the Information Environment (JCOIE)*. July 25, 2018. https://www.jcs.mil/Portals/36/Documents/Doctrine/concepts/joint_concepts_jcoie.pdf?ver=2018-08-01-142119-830.

Kagan, Robert. 2022. "The Price of Hegemony: Can America Learn to Use Its Power?" *Foreign Affairs* (May/June). https://www.foreignaffairs.com/articles/ukraine/2022-04-06/russia-ukraine-war-price-hegemony.

Kassam, Karim-Aly. 2021. "Those We Lost in Afghanistan Did Not Die in Vain." *The Hill*, September 5, 2021. https://thehill.com/opinion/international/570728-those-we-lost-in-afghanistan-did-not-die-in-vain/.

Kelly, John, Jim Sergent, and Donovan Slack. 2019. "Death Rates, Bedsores, ER Wait Times: Where Every VA Hospital Lags or Leads Other Medical Care." *USA Today*, December 16, 2019. https://www.usatoday.com/in-depth/news/investigations/2019/02/07/where-every-va-hospital-lags-leads-other-care/2511739002/.

Kelsey, Dick. 2003. "Today in Country Music History." *United Press International*, May 7, 2003. https://www.upi.com/Archives/2003/05/07/Country-Music-News/9381052280000/.

Kennedy, John F. 1961. "Inaugural Address." January 20, 1961. https://www.jfklibrary.org/learn/about-jfk/historic-speeches/inaugural-address.

Kepner, Timothy J. 2001. "Torture 101: The Case against the United States Atrocities Committed by the School of the Americas Alumni." *Penn State International Law Review* 19(3): 475–529.

Kershner, Seth, and Scott Harding. 2015. "Do Military Recruiters Belong in Schools?" *EducationWeek*, October 27, 2015. https://www.edweek.org/leadership/opinion-do-military-recruiters-belong-in-schools/2015/10.

King, Martin Luther, Jr. 1967. "Beyond Vietnam: A Time to Break Silence." Speech delivered at Manhattan's Riverside Church, April 4, 1967. https://www2.hawaii.edu/~freeman/courses/phil100/17.%20MLK%20Beyond%20Vietnam.pdf.

Kipling, Rudyard. 1899. "The White Man's Burden." https://www.kiplingsociety.co.uk/poem/poems_burden.htm.

Kissinger, Henry. 2003. *Ending the Vietnam War: A History of America's Involvement in and Extrication from the Vietnam War.* New York: Simon & Schuster.

Kitfield, James. 1996. "School for Scandal." *National Journal*, October 5, 1996.

Krebs, Ronald R., and Robert Ralston. 2022. "Patriotism or Paychecks: Who Believes What About Why Soldiers Serve." *Armed Forces and Society* 48 (1): 25–48.

Kreps, Sarah E. 2018. *Taxing Wars: The American Way of War Finance and the Decline of Democracy.* New York: Oxford University Press.

Kukathas, Chandran. 2003. *The Liberal Archipelago: A Theory of Diversity and Freedom.* New York: Oxford University Press.

Labonte, Marc, and Mindy Levit. 2008. *Financing Issues and Economic Effects of American Wars.* Congressional Research Service, July 29, 2008.

Lennard, Natasha. 2022. "The Real Life of Chelsea Manning." *Lux* 6 (Winter). https://lux-magazine.com/article/chelsea-manning-memoir/.

Lewis, Anthony. 1975. "The Kissinger Doctrine." *New York Times*, February 27, 1975. https://www.nytimes.com/1975/02/27/archives/the-kissinger-doctrine.html.

Linge, Mary Kay, and Maddie Panzer. 2022. "Watchdog Calls Out the Government's Most Ridiculous Spending." *New York Post*, July 2, 2022. https://nypost.com/2022/07/02/watchdog-calls-out-the-governments-most-ridiculous-spending/.

MacAskill, Ewen. 2003. "US Claims 45 Nations in 'Coalition of Willing.'" *The Guardian*, March 18, 2003. https://www.theguardian.com/world/2003/mar/19/iraq.usa.

Madison, James. 1865. "Political Observations, April 20, 1795." In *Letters and Other Writings of James Madison*, vol. 4, 485–505. Philadelphia: J. B. Lippincott & Co.

Malhotra, Anjana. 2005. "Witness to Abuse: Human Rights Abuses Under the Material Witness Law Since September 11." Human Rights Watch 17 (2): 1–71.

Marie, Samantha. 2018. "Lil Wayne Thanks 'Real Friend' Jay-Z for Helping to Pay Off Huge Tax Bill." NME, December 28, 2018. https://www.nme.com/news/music/lil-wayne-jay-z-tax-bill-2425114.

Marine Corps Directorate of Analytics & Performance Optimization. 2019. *Blast Overpressure Effects.* March 2019. https://www.hqmc.marines.mil/ Portals/61/Users/019/71/4371/Overpressure%20Study%20Report%20 20191025.pdf?ver=Nta6RKsuKvaHCTG_HrY1MQ%3D%3D.

Martinez, Michael. 2014. "Obama: 'Won't Be a Military Solution' if Iraqi Political Structure Not Fixed." *CNN Politics,* June 20, 2014. https://www. cnn.com/2014/06/20/politics/obama-iraq.

McClellan, Scott. 2008. *What Happened: Inside the Bush White House and Washington's Culture of Deception.* New York: Public Affairs Books.

McCloskey, Deirdre Nansen. 2019. *Why Liberalism Works: How True Liberal Values Produce a Freer, More Equal, Prosperous World for All.* New Haven, CT: Yale University Press.

McKinney, Kelsey. 2014. "How the CIA Used Music to 'Break' Detainees." *Vox,* December 11, 2014. https://www.vox.com/2014/12/11/7375961/ cia-torture-music.

McNulty, Mel. 1999. "Media Ethnicization and the International Response to War and Genocide in Rwanda." In *The Media of Conflict: War Reporting and Representations of Ethnic Violence,* edited by Tim Allen and Jean Seaton, 268–286. London: Zed Books.

Meet the Press, NBC News. 2002. "The Vice President Appears on Meet the Press with Tim Russert." September 8, 2002. https://georgewbush-whitehouse.archives.gov/vicepresident/news-speeches/speeches/ vp20010916.html.

Mencken, H. L. 1982. *A Mencken Chrestomathy.* New York: Vintage Books.

Meredith, Elizabeth T. 2010. "Jane Fonda: Repercussions of Her 1972 Visit to North Vietnam." Research report, Air Command and Staff College. https://apps.dtic.mil/sti/pdfs/AD1019243.pdf.

Mises, Ludwig von. 1996. *Liberalism: The Classical Tradition.* Irvington-on-Hudson, NY: The Foundation for Economic Education, Inc.

Miyares, Jason S. 2022. Letter to Secretary of Homeland Security Alejandro Mayorkas. May 5, 2022. https://content.govdelivery.com/attachments/ MTAG/2022/05/06/file_attachments/2152159/DGB%20Letter_Final.pdf.

Monboit, George. 2005. "The U.S. Used Chemical Weapons in Iraq—and Then Lied About It." *The Guardian,* November 14, 2005. https://www. theguardian.com/politics/2005/nov/15/usa.iraq.

Mongilio, Heather. 2022. "Latest Military Sexual Assault Report Shows 'Tragic' Rise in Cases, Pentagon Officials Say." *USNI News*, September 1, 2022. https://news.usni.org/2022/09/01/latest-military-sexual-assault-report-shows-tragic-rise-in-cases-pentagon-officials-say.

Montgomery, David. 2003. "The NFL's New Turf." *Washington Post*, September 1, 2003. https://www.washingtonpost.com/archive/lifestyle/2003/09/01/the-nfls-new-turf/fd3e1160-7b16-4b8d-a7cd-df1dbef3e06c/.

Morgan, Stokeley W. 1926. "Lecture Delivered Before the Foreign Service School, Department of State." Diplomatic Cable No. 462, January 29, 1926. National Archives.

Morris, David Z. 2022. "Deplatformed by PayPal, Antiwar Journalists Speak Out." *CoinDesk*, May 23, 2022. https://www.coindesk.com/layer2/2022/05/23/deplatformed-by-paypal-antiwar-journalists-speak-out/.

Morrison, Sarah. 2012. "Iraq Records Huge Rise in Birth Defects." *The Independent*, October 14, 2012. https://www.independent.co.uk/life-style/health-and-families/health-news/iraq-records-huge-rise-in-birth-defects-8210444.html.

Moses, Paul. 2010. "We Don't Do Body Counts." *Commonwealth*, October 24, 2010. https://www.commonwealmagazine.org/we-dont-do-body-counts.

Moss, Michael. 2010. "Getting to El Dorado Canyon: The Reagan Administration's 1986 Decision to Bomb Libya." *American Intelligence Journal* 28 (2): 45–49.

Mueller, John. 2021. *The Stupidity of War: American Foreign Policy and the Case for Complacency.* New York: Cambridge University Press.

———. 2006. *Overblown: How Politicians and the Terrorism Industry Inflate National Security Threats and Why We Believe Them.* New York: Simon & Schuster.

Mueller, John, and Mark G. Stewart. 2011. *Terror, Security, and Money: Balancing the Risks, Benefits, and Costs of Homeland Security.* New York: Oxford University Press.

Munshi, Neil. 2021. "How the Death of Gaddafi Is Still Being Felt by Libya's Neighbors." *Financial Times*, September 8, 2021. https://www.ft.com/content/bf3c06a2-3cff-4817-8c2d-519ca79f921f.

Myers, Meghann. 2023. "Experts, Data Point to Women as Best Military Recruiting Pool." *Military Times*, January 26, 2023. https://www.

militarytimes.com/news/your-military/2023/01/26/experts-data-point-to-women-as-best-military-recruiting-pool/.

Myre, Greg. 2017. "Once Reserved for Spies, Espionage Act Now Used Against Suspected Leakers." *NPR*, June 28, 2017. https://www.npr.org/sections/parallels/2017/06/28/534682231/once-reserved-for-spies-espionage-act-now-used-against-suspected-leakers.

National Park Service. 2007. *World War II and the American Home Front: A National Historic Landmarks Theme Study*. U.S. Department of the Interior. https://www.nps.gov/subjects/nationalhistoriclandmarks/upload/WWII_and_the_American_Home_Front-508.pdf.

Newman, Nic, Richard Fletcher, Craig T. Robertson, Kristen Eddy, and Kleis Nielsen. 2022. *Reuters Institute Digital News Report 2022*. Reuters Institute for the Study of Journalism. https://reutersinstitute.politics.ox.ac.uk/sites/default/files/2022-06/Digital_News-Report_2022.pdf.

Nixon Presidential Materials Project. 2001. "Oval Office Meeting with Bob Haldeman." *Nixon Tapes, Monday, 14 June 1971*. http://nsarchive2.gwu.edu/NSAEBB/NSAEBB48/oval.pdf.

Nossiter, Adam. 2012. "Qaddafi's Weapons, Taken by Old Allies, Reinvigorate an Insurgent Army in Mali." *New York Times*, February 5, 2012. https://www.nytimes.com/2012/02/06/world/africa/tuaregs-use-qaddafis-arms-for-rebellion-in-mali.html.

Obama, Barack. 2010. "Remarks by the President in Address to the Nation on the End of Combat Operations in Iraq." August 31, 2010. https://obamawhitehouse.archives.gov/the-press-office/2010/08/31/remarks-president-address-nation-end-combat-operations-iraq.

———. 2009. "Remarks by the President on a New Strategy for Afghanistan and Pakistan." March 27, 2009. https://obamawhitehouse.archives.gov/the-press-office/remarks-president-a-new-strategy-afghanistan-and-pakistan.

Orth, Taylor. 2022. "Airport Security: Despite the Inconvenience, Most Americans Say It's Effective." YouGov US, January 14, 2022. https://today.yougov.com/topics/travel/articles-reports/2022/01/14/airport-security-despite-inconvenience-most-americ.

Parry, Robert, and Norman Solomon. 2021. "Colin Powell: Military Adviser in Vietnam During My Lai Massacre." *Veterans for Peace,* October 20, 2021. https://www.vietnamfulldisclosure.org/colin-powell-my-lai-massacre/.

Paul, Christopher, and James Kim. 2004. *Reports on the Battlefield: The Embedded Press System in Historical Context.* RAND Corporation. https://www.rand.org/content/dam/rand/pubs/monographs/2004/RAND_MG200.pdf.

Pelosi, Nancy. 2022. "Transcript of Speaker Pelosi's Remarks at Weekly Press Conference." May 12, 2022. https://pelosi.house.gov/news/press-releases/transcript-of-speaker-pelosi-s-remarks-at-weekly-press-conference-10.

Pérez, Louis A., Jr. 1979. "Cuba Between Empires, 1898–1899." *Pacific Historical Review* 48 (4): 473–500.

Peter, Josh. 2020. "Professional Bull Riders Get Millions from Border Patrol for Patriotism and Plugs." *USA Today*, January 21, 2020. https://www.usatoday.com/story/sports/2020/01/21/immigration-sports-league-has-unique-partnership-border-patrol/2798296001/.

Pew Research Center. 2015. "Public Continues to Back U.S. Drone Attacks." May 28, 2015. https://www.pewresearch.org/politics/2015/05/28/public-continues-to-back-u-s-drone-attacks/.

———. 2003. "Embedded Reporters." April 3, 2003. https://www.pewresearch.org/journalism/2003/04/03/embedded-reporters/.

Philipps, Dave. 2023. "A Secret War, Strange New Wounds, and Silence from the Pentagon." *New York Times*, November 5, 2023. https://www.nytimes.com/2023/11/05/us/us-army-marines-artillery-isis-pentagon.html.

Piazza, J. A. 2013. "Regime Age and Terrorism: Are New Democracies Prone to Terrorism?" *International Interactions* 39 (2): 246–263.

Pilkington, Ed. 2013. "Declassified NSA Files Show Agency Spied on Muhammad Ali and MLK." *The Guardian*, September 26, 2013. https://www.theguardian.com/world/2013/sep/26/nsa-surveillance-anti-vietnam-muhammad-ali-mlk.

Place, Nathan. 2021. "Stephen Miller Reaches Back to 1972 to Accuse Jane Fonda of Treason." *The Independent*, June 10, 2021. https://www.independent.co.uk/news/world/americas/us-politics/stephen-miller-jane-fonda-treason-b1863524.html.

Population Reference Bureau. 2007. "Latinos Claim Larger Share of U.S. Military Personnel." *PRB*, October 13, 2007. https://www.prb.org/resources/latinos-claim-larger-share-of-u-s-military-personnel/.

Posner, Eric A., and Adrian Vermeule. 2007. *Terror in the Balance: Security, Liberty, and the Courts.* New York: Oxford University Press.

Priest, Dana, and William M. Arkin. 2011. *Top Secret America: The Rise of the New American Security State*. New York: Little, Brown, and Company.

Psaki, Jen. 2021. "Press Briefing by Press Secretary Jen Psaki and Surgeon General Dr. Vivek H. Murthy." July 15. https://www.whitehouse.gov/briefing-room/press-briefings/2021/07/15/press-briefing-by-press-secretary-jen-psaki-and-surgeon-general-dr-vivek-h-murthy-july-15-2021/.

Public Law 107-56. 2001. "Uniting and Strengthening America by Providing Appropriate Tools Required to Intercept and Obstruct Terrorism (USA PATRIOT ACT) Act of 2001."

Reagan, Ronald. 1986. "The President's News Conference." American Presidency Project, April 9, 1986. https://www.presidency.ucsb.edu/documents/the-presidents-news-conference-959.

Reifler, Jason, and Christopher Gelpi. 2006. "Casualties, Polls, and the Iraq War." *International Security* 31 (2): 186–198.

Rejali, Darius. 2007. *Torture and Democracy*. Princeton, NJ: Princeton University Press.

Richburg, Keith B. 1993. "Criticism Mounts over Somali Raid." *Washington Post*, July 15, 1993. https://www.washingtonpost.com/archive/politics/1993/07/15/criticism-mounts-over-somali-raid/2ff88b7b-bde2-48a4-a7ba-66dfb94312fa/.

Rippon, Jo. 2020. *The Art of Protest: A Visual History of Dissent and Resistance*. Watertown, MA: Charlesbridge Publishing.

Roberts, Lawrence. 2020. "Who Was Behind the Largest Mass Arrest in U.S. History?" *New York Times*, August 6, 2020. https://www.nytimes.com/2020/08/06/opinion/nixon-trump-protests-military.html.

Rockoff, Hugh. 2012. *America's Economic Way of War: War and the US Economy from the Spanish-American War to the Persian Gulf War*. Cambridge, UK: Cambridge University Press.

Rohter, Larry. 2002. "Argentina Charges Ex-Dictator and Others in 'Dirty War' Deaths." *New York Times*, July 11, 2002.

Roosevelt, Franklin D. 1942a. "Annual Budget Message." National Archives, January 5, 1942. https://www.presidency.ucsb.edu/documents/annual-budget-message-2.

———. 1942b. "Executive Order 9066—Authorizing the Secretary of War to Prescribe Military Areas." National Archives, February 19, 1942. https://www.archives.gov/milestone-documents/executive-order-9066#transcript.

_____.1942c. "Executive Order 9182—Consolidating Certain War Information Functions into an Office of War Information." American Presidency Project, June 13, 1942. https://www.presidency.ucsb.edu/documents/executive-order-9182-consolidating-certain-war-information-functions-into-office-war.

Satter, Raphael. 2020. "U.S. Court: Mass Surveillance Program Exposed by Snowden Was Illegal." *Reuters*, September 2, 2020. https://www.reuters.com/article/us-usa-nsaspying/u-s-court-mass-surveillance-program-exposed-by-snowden-was-illegalidUSKBN25T3CK.

Savage, Charlie. 2023. "Pentagon Blocks Sharing Evidence of Possible Russian War Crimes with Hague Court." *New York Times*, March 8, 2023. https://www.nytimes.com/2023/03/08/us/politics/pentagon-war-crimes-hague.html.

Savell, Stephanie. 2021. "United States Counterterrorism Operations 2018–2020." Watson Institute for International and Public Affairs. https://watson.brown.edu/costsofwar/papers/2021/USCounterterrorismOperations.

Schell, Jonathan. 2004. *The Unconquerable World: Power, Nonviolence, and the Will of the People*. New York: Henry Holt and Company.

Schmitt, Eric, and Carolyn Marshall. 2006. "In Secret Unit's 'Black Room,' a Grim Portrait of U.S. Abuse." *New York Times*, March 19, 2006. https://www.nytimes.com/2006/03/19/world/middleeast/in-secret-units-black-room-a-grim-portrait-of-us-abuse.html.

Scott, Douglas D. 2001. "Firearms Identification in Support of Identifying a Mass Execution at El Mozote, El Salvador." *Historical Archaeology* 35 (1): 79–86.

Seligman, Edwin R. 1918. "The War Revenue Act." *Political Science Quarterly* 33 (1): 1–37.

Senate Select Committee on Intelligence. 2014. *Central Intelligence Agency's Detention and Interrogation Program*. Washington, DC: Government Printing Office.

Shafer, Jack. 2003. "The PR War." *Slate*, March 25, 2003. https://slate.com/news-and-politics/2003/03/the-pr-war.html.

Shakespeare, William. 1600. *The Merchant of Venice*, edited by Barbara A. Mowat and Paul Werstine. Folger Shakespeare Library. https://folger-main-site-assets.s3.amazonaws.com/uploads/2022/11/the-merchant-of-venice_PDF_FolgerShakespeare.pdf.

Shane, Scott. 2015. "Anwar al-Awlaki, Yemen, and American Counterterrorism Policy." Brookings, September 17, 2015. https://www.brookings.edu/events/anwar-al-awlaki-yemen-and-american-counterterrorism-policy/.

Sharp, Gene. 1990. *Civilian-Based Defense: A Post-Military Weapons System.* Princeton, NJ: Princeton University Press.

Shearer, Elisa. 2021. "More than Eight-in-Ten Americans Get News from Digital Devices." Pew Research Center, January 12. https://www.pewresearch.org/short-reads/2021/01/12/more-than-eight-in-ten-americans-get-news-from-digital-devices/.

Shepp, Jonah. 2023. "Don't Blame Gazans for Hamas." *New York: Intelligencer,* October 22, 2023. https://nymag.com/intelligencer/2023/10/dont-blame-gazans-for-hamas.html

Shoker, Sarah. 2018. "Military-Age Males in U.S. Counterinsurgency and Drone Warfare." PhD dissertation, McMaster University. https://macsphere.mcmaster.ca/bitstream/11375/24294/2/Shoker_Sarah_2018April_PhD.pdf.

Singh, Abhishek Kumar. 2022. "The North Korean Threat: A Thorn in Seoul's Side." Observer Research Foundation, December 6. https://www.orfonline.org/expert-speak/the-north-korean-threat/.

Siripurapu, Anshu. 2021. "What Is the Defense Production Act?" Council on Foreign Relations. https://www.cfr.org/in-brief/what-defense-production-act.

Smith, Clive Stafford. 2008. "Welcome to 'the Disco.'" *The Guardian,* June 18, 2008. https://www.theguardian.com/world/2008/jun/19/usa.guantanamo.

Smith, David. 2021. "'She Was Very Complicated. She Was a Conundrum': Who Was the Real Lucille Ball?" *The Guardian,* December 8. https://www.theguardian.com/tv-and-radio/2021/dec/07/she-was-very-complicated-she-was-a-conundrum-who-was-the-real-lucille-ball.

Smith, Grady. 2015. "Is Country Music Ready to Forgive the Dixie Chicks?" *The Guardian,* November 19. https://www.theguardian.com/music/2015/nov/19/the-dixie-chicks-tour-is-country-music-ready-to-forgive.

Smith, Matthew. 2018. "Are the Troops Heroes? Americans, Britons and Germans Feel Very Differently." YouGov US, September 26. https://today.yougov.com/politics/articles/21617-are-troops-heroes-americans-britons-and-germans-fe-1.

Smith, Noah. 2022. "The War Economy: An Organizing Principle for the Next 20 Years of U.S. Political Economy." *Noahpinion*, July 14, 2022. https://noahpinion.substack.com/p/the-war-economy.

Snapes, Laura. 2020. "The Chicks: 'We Were Used and Abused by Everybody Who Wanted to Make Money Off Us.'" *The Guardian*, July 18. https://www.theguardian.com/music/2020/jul/18/dixie-chicks-used-and-abused-by-everybody-who-wanted-to-make-money-off-us.

Stone, Oliver, and Peter Kuznick. 2012. *The Untold Story of the United States.* New York: Gallery Books.

Sultan, Ahmad, and Abdul Qadir Sediqi. 2019. "U.S. Drone Strike Kills 30 Pine Nut Farm Workers in Afghanistan." *Reuters*, September 19, 2019. https://www.reuters.com/article/us-afghanistan-attack-drones/u-s-drone-strike-kills-30-pine-nut-farm-workers-in-afghanistan-idUSKBN1W40NW.

The Guardian editors. 2003. "France and Germany Unite Against Iraq War." *The Guardian*, January 22, 2003. https://www.theguardian.com/world/2003/jan/22/germany.france.

Thorndike, Joseph J. 2022. "Timelines in Tax History: From 'Class Tax' to 'Mass Tax' During World War II." *Tax Notes*, September 19, 2022. https://www.taxnotes.com/tax-history-project/timelines-tax-history-class-tax-mass-tax-during-world-war-ii/2022/09/16/7f3s2.

Thrall, A. Trevor, and Jane K. Cramer, eds. 2009. *American Foreign Policy and the Politics of Fear: Threat Inflation Since 9/11*. New York: Routledge.

Time editors. 1942. "Taxes: Bigger and Better." *Time*, October 19, 1942. https://time.com/vault/issue/1942-10-19/page/23/.

Tirman, John. 2011. *The Deaths of Others: The Fate of Civilians in America's Wars*. New York: Oxford University Press.

Trump, Donald. 2020. "Memorandum on Order Under the Defense Production Act Regarding 3M Company." April 2, 2020. https://trumpwhitehouse.archives.gov/presidential-actions/memorandum-order-defense-production-act-regarding-3m-company/.

Turner, Henry A. 1957. "Woodrow Wilson and Public Opinion." *Public Opinion Quarterly*, 21 (4): 505–520.

Twight, Charlotte. 1975. *America's Emerging Fascist Economy*. New York: Arlington House Publishers.

United Nations Office for Disarmament Affairs. n.d. "Arms Trade." https://www.un.org/disarmament/convarms/att/.

United Nations Office of the High Commissioner for Human Rights. 2022. "U.N. Human Rights Office Estimates More than 306,000 Civilians Were Killed over 10 Years in Syria." June 28, 2022. https://www.ohchr.org/en/press-releases/2022/06/un-human-rights-office-estimates-more-306000-civilians-were-killed-over-10.

Unknown Author. 2021a. Email. May 28, 2021. http://web.archive.org/web/20220907150818/https://ago.mo.gov/docs/default-source/press-releases/free-speech-pitch-thread-docs/hhs-fb-exhibit.pdf?sfvrsn=55bd83df_2.

Unknown Author. 2021b. Email. July 19, 2021. http://web.archive.org/web/20220911210034/https://ago.mo.gov/docs/default-source/press-releases/free-speech-pitch-thread-docs/hhs-fb-email-1.pdf?sfvrsn=53bc4454_2.

Unknown Author. 2021c. Email. July 28, 2021. http://web.archive.org/web/20221102181614/https://ago.mo.gov/docs/default-source/press-releases/free-speech-pitch-thread-docs/cdc-fb-monthly-debunk.pdf?sfvrsn=3508a21f_2.

U.S. Army. 2022. *FM 3-0 Operations*. https://armypubs.army.mil/epubs/DR_pubs/DR_a/ARN36290-FM_3-0-000-WEB-2.pdf.

U.S. Army Junior ROTC 2023. "Army Junior ROTC Program Overview." https://www.usarmyjrotc.com/army-junior-rotc-program-overview/.

U.S. Department of Defense. 2022. *Military and Security Developments Involving the People's Republic of China*. https://media.defense.gov/2022/Nov/29/2003122279/-1/-1/1/2022-MILITARY-AND-SECURITY-DEVELOPMENTS-INVOLVING-THE-PEOPLES-REPUBLIC-OF-CHINA.PDF.

U.S. Department of Homeland Security. 2022a. "Fact Sheet: DHS Internal Working Group Protects Free Speech and Other Fundamental Rights When Addressing Disinformation That Threatens the Security of the United States." May 2, 2022. https://www.dhs.gov/news/2022/05/02/fact-sheet-dhs-internal-working-group-protects-free-speech-other-fundamental-rights.

————. 2022b. *National Terrorism Advisory System Bulletin*, May 24, 2022. https://www.dhs.gov/sites/default/files/ntas/alerts/22_1130_S1_NTAS-Bulletin-508.pdf.

U.S. Department of Justice. 2022. "Member of Iran's Islamic Revolution Guard Corps (IRGC) Charged with Plot to Murder the Former National Security Advisor." August 10, 2022. https://www.justice.gov/opa/pr/member-irans-islamic-revolutionary-guard-corps-irgc-charged-plot-murder-former-national.

U.S. Department of State. 2021. "The Chinese Communist Party: Threatening Global Peace and Security." https://2017-2021.state.gov/wp-content/uploads/2020/10/FINAL20one-pager20Threatening20Global20Peace20Security-1.pdf.

U.S. Department of Veterans Affairs. 2021. *Department of Veterans Affairs (VA) Board of Veterans' Appeals Annual Report Fiscal Year (FY) 2020*. https://www.bva.va.gov/docs/Chairmans_Annual_Rpts/BVA2020AR.pdf.

U.S. Government Printing Office. 1953. "Lucille Ball HUAC Testimony." https://learn.k20center.ou.edu/lesson/2348/Lucille%20Ball%20HUAC%20Testimony-%20HUAC%20Versus%20Hollywood.pdf?rev=18308.

U.S. House of Representatives Committee on Government Reform. 2004. *Iraq on the Records: The Bush Administration's Public Statements on Iraq*. Minority Staff Special Investigations Division, March 16, 2004. https://www.cs.cornell.edu/gries/howbushoperates/pdf_admin_iraq.pdf.

U.S. Office of War Information. 1942. *Government Information Manual for the Motion Picture Industry*. Washington, DC: Office of War Information.

Ware, Doug. 2023. "Reports of Sexual Assaults Increased in Navy, Air Force and Marines in 2022; Army Saw Decline." *Stars and Stripes*, April 27, 2023. https://www.stripes.com/theaters/us/2023-04-27/military-sexual-assaults-report-9940513.html.

Washington Post editors. 2003. "Washington Post Poll: Saddam Hussein and the Sept. 11 Attacks." *Washington Post*, September 6, 2003. https://www.washingtonpost.com/wp-srv/politics/polls/vault/stories/data082303.htm.

Watson Institute for International and Public Affairs at Brown University. 2022a. "Afghan Civilians." *Costs of War*. https://watson.brown.edu/costsofwar/costs/human/civilians/afghan.

_____. 2022b. "Estimate of U.S. Post-9/11 Spending, in $ Billions FY2001–FY2022." *Costs of War.* https://watson.brown.edu/costsofwar/figures/2021/BudgetaryCosts.

Watt, J. J. 2021. X post, March 23, 2021. https://twitter.com/JJWatt/status/1367112474722525191.

Wezeman, Peter D., Alexandra Kuimova, and Siemon T. Wezeman. 2022. "Trends in International Arms Transfers, 2021." *SIPRI Fact Sheet*, March 2022. Stockholm International Peace Research Institute. https://www.sipri.org/sites/default/files/2022-03/fs_2203_at_2021.pdf.

White, William Allen. 1937. *Forty Years on Main Street.* New York: Farrar & Rinehart.

White House. 2023a. "Press Briefing by Press Secretary Karine Jean-Pierre and NSC Coordinator for Strategic Communications John Kirby." October 12, 2023. https://www.whitehouse.gov/briefing-room/press-briefings/2023/10/12/press-briefing-by-press-secretary-karine-jean-pierre-and-nsc-coordinator-for-strategic-communications-john-kirby-21/.

_____. 2023b. "Remarks by Vice President Harris in Press Gaggle, Bletchley, United Kingdom." November 2, 2023. https://www.whitehouse.gov/briefing-room/speeches-remarks/2023/11/02/remarks-by-vice-president-harris-in-press-gaggle-bletchley-united-kingdom/.

Whitlock, Craig. 2021. *The Afghanistan Papers: A Secret History of the War.* New York: Simon & Schuster.

Wilson, Julie. 2017. "Here's Why TSA Agents Make You Remove Your Shoes at the Airport." *ABC 11*, October 6. https://abc11.com/why-do-i-take-my-shoes-off-at-the-airport-what-can-on-carry--rdu-flights/2499196/.

Wilson, Woodrow. 1918. "Letter to Bernard M. Baruch Requesting Acceptance of Nomination as Chair of the War Industries Board." American Presidency Project, March 4, 1918. https://www.presidency.ucsb.edu/documents/letter-bernard-m-baruch-requesting-acceptance-nomination-chair-the-war-industries-board.

_____. 1917a. "Joint Address to Congress Leading to a Declaration of War Against Germany." National Archives, April 2, 1917. https://www.archives.gov/milestone-documents/address-to-congress-declaration-of-war-against-germany.

_____. 1917b. "Proclamation 1364." April 6, 1917. https://millercenter.org/the-presidency/presidential-speeches/april-6-1917-proclamation-1364.

Winkler, Allan. 1978. *The Politics of Propaganda: The Office of War Information, 1942–1945.* New Haven, CT: Yale University Press.

Wintour, Patrick, and Jon Henley. 2003. "Don't Blame Us for Conflict, Protest French." *The Guardian*, March 19, 2023. https://www.theguardian.com/world/2003/mar/20/politics.iraq/.

Wong, May. 2019. "Stanford Research Finds Flaws in Veterans' Claims System." *Stanford News*, March 6, 2019. https://news.stanford.edu/2019/03/06/new-research-finds-flaws-veterans-claims-system/.

World Bank. 2022. "Rwanda." https://data.worldbank.org/country/rwanda.

X. 2021a. Post by user @jonfrensley2021, March 4, 2021. https://twitter.com/jonfrensley2021/status/1367496249314848780.

————. 2021b. Post by user @Tricia5618, March 5, 2021. https://twitter.com/Tricia5618/status/1367742675613593602.

————. 2021c. Post by user @PalimenoForGAD1, March 26, 2021. https://twitter.com/PalimenoForGAD1/status/1368188679773110286.

Yang, Patricia J., Morgan LaMarca, Candice Kaminski, Daniel I. Chu, and David L. Hu. 2017. "Hydrodynamics of Defecation." *Soft Matter* 13 (29): 4960–4970.

Yayboke, Erol. 2022. "Update on Forced Displacement Around Ukraine." Center for Strategic and International Studies, October 3, 2022. https://www.csis.org/analysis/update-forced-displacement-around-ukraine.

Yeung, Douglas, Christina E. Steine, Charitra M. Hardison, Lawrence M. Hanser, and Kristy N. Kamarck. 2017. "Recruiting Policies and Practices for Women in the Military: Views from the Field." https://www.rand.org/pubs/research_reports/RR1538.html.

Zavadski, Katie. 2016. "The Terrorists Drones Couldn't Silence." *Politico*, September 30, 2016. https://www.politico.com/magazine/story/2016/09/samir-khan-drone-strike-anwar-al-awlaki-214308/.

Zwirko, Colin, and Jeongmin Kim. 2022. "Kim Jong Un Says He Will 'Never Give Up' Nuclear Weapons, Rejects Future Talks." *NK News*, September 9, 2022. https://www.nknews.org/2022/09/kim-jong-un-says-he-will-never-give-up-nuclear-weapons-rejects-future-talks/.

3M. n.d. "3M Response to Defense Production Act Order." https://news.3m.com/2020-04-03-3M-Response-to-Defense-Production-Act-Order.

Index

A

Abu Ghraib prison, 50, 116, 118

activist organizations, 69–70

Adams, John, 99–100

Addams, Jane, 144

Afghanistan: arms exports to, 123; casualties in, 49, 51, 62; impediments to democracy in, 135; Tillman's death in, 52–53; U.S. activities in, 55, 104; withdrawal from, 12, 55, 128, 135

Africa, 112

Al-Aulaqi, Abdulrahman, 70–71

Al-Aulaqi, Anwar, 70

Algeria, 127

Ali, Muhammad, 104

Alien and Sedition Acts, 99–100

Alien Enemies Act, 100

Allende, Salvador, 133

Al-Qaeda, 33, 34, 36, 49, 70

American Civil Liberties Union (ACLU), 68, 70

antiwar activism, 105, 106–107

antiwar protests, 67, 97

Argentina, 121

arms exports, 122–123, 142

Arms Trade Treaty (ATT), 121

Army Corps of Engineers, 78

Arnaz, Desiderio "Desi," 66

assassination, 121

asset seizure, 69–70

Atlácatl Battalion, 122

Atta, Mohamed, 34

audio torture, 117–118

B

Baldor, Lolita, 59

Ball, Lucille, 66–67

Barrios, José Domingo Monterrosa, 122

Barstow, David, 38

Beckley, Michael, 3

Berlin, Irving, 88–89, 90

Biden, Joe: on democracy, 6; executive orders, 80; on Facebook and the pandemic, 21; on U.S. withdrawal from Afghanistan, 128, 135

Biden administration: and mis/disinformation, 43–44; pandemic policies of, 40; on Russia and Ukraine, 9, 97; on the Tower 22 attack, 141

bin Laden, Osama, 34

Birth of a Nation (film), 26

blackmail, 120, 121

Black Market, 74

Black Panthers, 104

Blair, Tony, 36–37

Boeing, 73

Bolívar, Simón, 133–134

Bolivia, 121

Bolton, John, 9

Brands, Hal, 3

Buchanan, James, 145

Bureau of Intelligence and Research, 35

Bush (George W.) administration:
communication strategies, 31, 33, 36,
38, 49; and the film industry, 27; on
Tillman's death, 52–53; and the war in
Iraq, 129–130
Bush, George W.: on Afghanistan, 135;
on Abu Ghraib, 50; on Iraq, 130;
meeting with Blair, 36–37; on military
casualties, 54; on WMDs, 35, 36
Butler, Smedley, 144

C
Calderón, Hernán Vildoso, 121
Cambodia, 103, 129
Cameron, David, 132
Camp Nama, 117
Canada, 12
capitalism, 73, 80, 81, 83, 140
Cappella, Rosella, 91
Carter, James Earl "Jimmy," 122
Castro, Fidel, 104, 111
casualties of war: military, 51–60;
military-age males (MAMs), 48–49;
minimizing, 47–48, 62; physical and
psychological, 55–57; sunk cost fallacy,
53–55; targeted killings, 70; *see also*
civilian casualties
Cavanaugh, Matt, 55
Cavin, Dennis D., 59
celebrity activism, 106–107
censorship, 114; by proxy, 109
Centers for Disease Control and
Prevention (CDC), 41
Chad, 127
chemical weapons, 119
Cheney, Dick, 33, 34, 35
Chile, 133
China: arms exports by, 122;
disinformation spread by, 21;
environmental impact of, 8; human
rights violations by, 7–8; military
intervention by, 7; use of propaganda
by, 7; as threat to global order, 6–7
Chinese Communist Party (CCP), 6–7
Chirac, Jacques, 131
Chomsky, Noam, 12

Churchill, Winston, 17
civilian casualties: in Afghanistan, 51,
104; birth defects and miscarriages, 51,
119–120; controlling reports of, 47–49;
credible reports of, 51; in Iraq, 104;
propaganda surrounding, 51; targeted
killings, 70–71; women and children,
49–50
civil rights activism, 105
Civil Service Reform Act (1978), 103
Clinton, William "Bill," 113
Clinton administration, 35, 116
coalition of the willing, 37, 131
Cohn, Lindsay, 60
COINTELPRO (Counterintelligence
Program), 104, 105
Cold War, 32, 66, 77, 122
Collins, John, 25
Committee on Public Information (CPI),
19–20, 26, 28, 32
Communist Party, 66
Congressional Budget Office, 89–90
Congressional Research Service, 91, 94
Consortium, 108
corporate greed, 94–95
COVID-19 pandemic, 21, 40, 78–79, 80,
94
Creel Committee, 19–20
crimes against humanity, 121
Cuba, 12, 104, 111–112, 113; Guantanamo
Bay, 116
currency expansion, 94–95

D
debt financing, 90–95, 140
defense contractors, 73–74, 78–79, 81
Defense Production Act (DPA), 77–81
democracy: American, 146; appreciation
of, 132; constitutional, 146; defense of,
6; gift of, 140; impediments to, 134–135;
in Iraq, 135; promotion of, 131; spread of,
12; threats to, 6
Democratic Republic of the Congo (Zaire),
112
Department of Defense, 25, 78, 102, 117,
120; and the film industry, 26–28; and

professional sports, 25–26, 29
Department of Health and Human
   Services (HHS), 41
Department of Homeland Security (DHS),
   11, 21, 22
Department of Justice, 67–68
diplomacy, 143; "olive branch," 18
disinformation, 40, 43–44
Disinformation Governance Board
   (DGB), 21, 22, 23
dissent: discrediting, 104; domestic, 104;
   preventing, 98; silencing, 97–98,
   100–101, 110
Dixie Chicks, 107
domino principle, 13
draft, opposition to, 101
drone strikes, 49, 70, 71, 141
due process of law, 110

E
Ecuador, 121
Egypt, 127
Einstein, Albert, 144
Eisenhower, Dwight D., 13, 81
electrocution, 117
Elizabethtown (film), 28
Ellsberg, Daniel, 102–103
El Mozote massacre, 122–123
El Salvador, 122
environmental issues, 8
Espionage Act (1917), 101, 103, 104
executions, 120
Executive Order 9066, 65
exhaustion exercises, 118
extortion, 120
extrajudicial killings, 70–71

F
Facebook, 40–41, 43
failure: emphasizing the future, 127–128;
   mitigating, 125–126; shifting blame,
   129–136
Fang, Lee, 42
fascism, military-economic, 82
fear, 67, 72, 124, 145
Federal Air Marshal Service (FAMS), 24

Federal Bureau of Investigation (FBI), 34,
   78, 104, 105
Federal Emergency Management Agency
   (FEMA), 78
Federal Trade Commission (FTC),
   108–109
Ferguson, Niall, 113
Few Good Men, A (film), 110
fiat currency, 94–95
Filipino people, 113, 115
film industry, 26–28
Fish, Hamilton, 76
Fleischer, Ari, 37
Fonda, Jane, 106–107
foreign policy: and the control
   of information, 19; of foreign
   governments, 6; funding for, 93–95, 130;
   proactive, 3; and public narrative, 4, 5,
   17; and sporting events, 25–26; support
   for, 31
Four Minute Men, 19
France, 122, 131
Franks, Tommy, 62
freedom: appreciation of, 132; of assembly,
   67, 110; of association, 64; bottom-up
   approach to, 143; economic, 83; and
   information, 23; loss of, 68, 71; of
   movement, 72; of the press, 44–45;
   promotion of, 131, 139–140; protection
   of, 44, 81, 145; of religion, 110; sacrifices
   associated with, 94; of speech, 44–45,
   64, 72, 101, 110; threats to, 6; *see also*
   liberty
free market, 73, 75, 81, 83, 140

G
Gaddafi, Muammar, 125–127
Galtieri, Leopoldo, 121
General Dynamics, 73
General Electric, 78–79
General Motors, 78–79
genocide, 112–113, 121
Germany, 85, 122
Gibbs, Robert, 71
Glick, Brian, 104
Government Information Manual for the

Motion Picture Industry (GIMMPI), 26–27
Grabell, Michael, 24
Great Society, 91
greenhouse gases, 8
Greenpeace, 69
Grenada, 131
Griffith, D. W., 26
Guantanamo Bay, 116
Guatemala, 119
Gulf War, 85

H
Haldeman, H. R., 102
Hamas, 5, 10, 43–44, 112
Harris, Kamala, 43
Hawley, Josh, 22
Hemingway, Ernest, 104
Hetfield, James, 118
Heymann, Phillip, 116
Higgs, Robert, 82–83
Hitchcock, Curtis N., 75–76
Hobbes, Thomas, 13
House Un-American Activities Committee (HUAC), 66
human rights: in China, 7–8; in Syria, 8–9; training in, 120, 121; in the U.S., 8; violations of, 133, 140
Hunt, Marsha, 66–67
Hussein, Saddam, 33, 35–36; and Al-Qaeda, 33–34, 36, 38

I
imperialism, 106; militaristic, 142
individualism, 94
Indochina, 13, 131
information: asymmetries of, 23–24, 101–102; control of, 19, 21–24, 28–29, 31, 40, 47–48; public health, 21; role of, 17–18; suppression of, 22; during WWI and WWII, 19–20
Instagram, 40
Internal Revenue Service, 89
International Atomic Energy Agency, 35
International Covenant on Civil and Political Rights (1966), 114

International Criminal Court in The Hague, 97
international law, 111, 112, 120, 121–122, 140
interrogation methods, 116–117, 120; *see also* torture
Iran, 9–10, 21, 141
Iraq: Abu Ghraib prison, 50, 116, 118; arms exports to, 123; Bush's statements on, 35; casualties in, 49; coalition of the willing in, 37, 131; impediments to democracy in, 135; U.S. invasion of, 12, 33, 37, 104; U.S. military in Baghdad, 117; U.S. occupation of, 141; U.S.; and Tower 22, 141; weapons of mass destruction in, 34–36, 37, 38
Iraq War: chemical weapons used in, 119; embedding media in, 39; end of, 128; funding for, 130; opposition to, 37, 107; portrayal of, 27, 33, 36, 38; support for, 37; thermobaric weapons in, 51, 119
Islamic extremism, 10
Islamic Revolutionary Guard Corps, 9
Islamic State (IS), 141
isolation, 116
Israel, 5, 10, 43–44, 112, 142

J
Jankowicz, Nina, 22, 23
Japan, 11, 85
Japanese internment, 65–66
Jefferson, Thomas, 11–12
Johnson, Jay L., 61–62
Johnson, Lyndon B., 91
Johnson administration, 91–92, 102
Joint Chiefs of Staff, 17
Joint Concept for Operating in the Information Environment (JCOIE), 17
Jones, James L., 61
Jordan, 141
journalists: embedded, 39–40, 108; hostile, 108; silencing, 108; *see also* media
Junior Reserve Officers' Training Corps (JROTC), 59, 61–62
Justice Department, 118

K
Kagan, Robert, 4
Kennedy, John F., 18
Khan, Samir, 70
Khmer Rouge, 129
Kim Jong Un, 10
King, Martin Luther Jr., 105
Kipling, Rudyard, 113, 115
Kirby, John, 43
Kissinger, Henry, 32, 102–103, 129, 133
Korean War, 89
Kristol, Irving, 22
Kuehl, Dan, 117–118

L
Laos, 129
Latin America, 132–133, 134; *see also* South
  America
Latin American military, 120
Lehrer, Jim, 35
Leninism, 133
Lewis, D. J., 117
liberalism, 44, 85, 116, 145; global, 125
Liberia, 113
liberty: enemies of, 142; fear of, 145; and
  information, 23; loss of, 70; promotion
  of, 139–140; protection of, 44, 81, 110,
  145; rhetoric of, 146; sacrifices associated
  with, 94; sacrificing, 63–66; vs. safety,
  72; *see also* information
Libya, 125–127, 131, 134, 141
Lisicki, George, 130
Lloyd-La Follette Act (1912), 103
Lochhead, David, 127
Lockheed Martin, 73
Logan Act, 106

M
MacArthur, Douglas, 145
Madison, James, 11, 142
Maines, Natalie, 107
Major League Baseball (MLB), 26
Major League Soccer (MLS), 26
Mali, 127
Manning, Chelsea, 104
Marxism, 133

material witness laws, 67–68
May Day Tribe, 67
Mayorkas, Alejandro, 22
McClellan, Scott, 31
McNulty, Mel, 112, 113
media: in conflict zones, 39; control of, 31,
  140; electronic devices, 32; embedding,
  39–40, 108; hostile, 108; print
  publications, 32; providing "objective"
  information, 38–39; public trust in, 32;
  see also journalists
Mencken, H. L., 146–147
Middle East: arms exports to, 122;
  impediments to democracy in, 134–135
military-age males (MAMs), 48–49
military-industrial complex (MIC), 81
military-industrial-congressional-complex
  (MICC), 81
military intervention: in Africa, 113; as
  cause of further violence, 140; by
  China, 7; cost of, 86–95; elements
  necessary for success, 114; funding
  for, 140; in Grenada, 131; in Jordan,
  141; in Libya, 125–127, 131–132, 134,
  141; problematic variables, 143; public
  support for, 52; sacrifices associated
  with, 47–49; *see also* U.S. military
Miller, Stephen, 107
MintPress, 108
misinformation, 21, 23, 43–44; on social
  media, 40–43
Miyares, Jason, 22
monetization of debt, 94–95
moral torture, 118
Morgan, Stokeley W., 4, 132, 133
Morris, David, 108
Motion Picture Association of America
  (MPAA), 27
Musk, Elon, 108–109
My Lai massacre, 49–50

N
napalm, 119
National Association for Stock Car Auto
  Racing (NASCAR), 25
national defense: and the Defense

Production Act, 77–81; and public goods, 82

National Defense Act (1916), 61

National Football League (NFL), 25, 26, 52, 53

National Hockey League (NHL), 25–26

National School Service, 19

national security: common people and, 142; information on, 23, 24; and international law, 111; paternalistic approach to, 143; scope of, 83; threats to, 21

National Security Agency, 90

National Security Council, 43

National Security Strategy, 6

National Terrorism Advisory System Bulletin, 11

Nation of Islam, 104

NATO, 126

New Deal, 87

Niger, 127

1984 (Orwell), 22

Nixon, Richard M., 102, 103

Nixon administration, 67

Noriega, Manuel Antonio, 121

North Korea, 10–11

Northrop Grumman, 73

nuclear weapons, 10

O

Obama, Barack: and the DPA, 78; on Iraq, 128; on Libya, 126, 127, 132; on the Middle East, 134–135; on military casualties, 55; on the role of the U.S., 13

Obama administration, 49, 103

Obey, David R., 93

Office of War Information (OWI), 20, 26, 27, 28, 32

Official Bulletin (CPI newspaper), 19

Oliver, Mari, 98

Operation Rescue, 69

organized crime, 122

Orwell, George, 22

ostracism, 99

P

pacifism, 12, 18, 97

Pakistan, 55

Palestine, 112, 142

Palestine Islamic Jihad (PIJ), 10

Panama, 121

Panama Canal, 131

PayPal, 108

peace: conditions necessary for, 143–144; liberal, 145

Pelosi, Nancy, 95

Pentagon: 9/11 attack on, 11; affiliations with social media, 42; on Russian war crimes, 97; *see also* U.S. military

Pentagon Papers, 102

People's Republic of China (PRC), 6–7

Philippines, 113, 115

Pinochet, Augusto, 133

Pledge of Allegiance, 98–99

Plumbers, 103–104

Pompeo, Mike, 9

Powell, Colin, 33, 35, 37, 48

price controls, 74, 75–76

privacy issues, 23, 72

private ownership, 82

private-public partnerships, 73, 83

Professional Bull Riders (PBR), 26

propaganda: Chinese, 7; from foreign sources, 51; government, 140

Psaki, Jen, 40

psychological warfare, 104

public goods, 82, 155n24

Putin, Vladimir, 3–4, 9; *see also* Russia

Q

Qadhafi, Muammar, 125–127

R

Ramsay, Gordon, 28

Rankin, Jeannette, 12

rationing, 74

Reagan, Ronald, 122, 126

Reid, Harry, 130

Rejali, Darius, 116

Revenue Act (1942), 87

Rice, Condoleezza, 33

Rodríguez, Guillermo "Bombita," 121
Roosevelt, Franklin D., 65; New Deal, 87;
    during WWII, 89
Rove, Karl, 27
Rumsfeld, Donald, 33, 102
Russia: arms exports by, 122; and Crimea,
    9; disinformation spread by, 21; and
    Iran, 9–10; and Syria, 8–9; as threat, 6,
    8–9; and Ukraine, 3–4, 5, 10, 97, 108,
    128, 141
Rwanda, 112–113

S
Saudi Arabia, 122
Savage, Charlie, 98
Schenck, Charles T., 101
School of the Americas (SOA), 120
Scott, Douglas, 123
Secretary of Homeland Security, 22
Sedition Act (1918), 101
self-determination, 146
self-governance, 110, 111, 116
sexual assault, 60–61
Shafer, Jack, 40
Shafter, William R., 111, 113
Share the Sacrifice Act, 93
Shoker, Sarah, 49
sleep deprivation, 116, 117
Small Arms Survey, 127
Snapes, Laura, 107
socialism, 22, 104; Islamic, 126
social media, 20–21; disinformation spread
    by, 40–43; as news source, 40
social shaming, 99
solar panels, 80
Somalia, 113
South America, 134; *see also* Latin America
Southeast Asia, 13, 129
South Korea, 10–11, 85
sovereignty, 112
Soviet Union, 122; *see also* Russia
Spanish-American War, 85, 111
sporting events, 24–25, 29
State Department, 6, 7, 8, 108; Division of
    Latin American Affairs, 132
Stolen Valor Act (2013), 56–57

Suárez, Hugo Banzer, 121
Sudan, 127
sunk cost fallacy, 53–55
surveillance operations, 67, 109; analysis of
    private data, 71–72
Syria, 8–9, 141

T
Taiwan, 11
Taliban, 12, 51, 55, 135
taxation: growth in revenue, 89–90; to
    support the military, 86–89; War Excise
    Taxes, 87
television shows, 27, 28
terrorism: 9/11 attacks, 11, 33, 34, 52, 55,
    141; Christmas Day (Detroit 2009), 70;
    and the Defense Production Act, 78;
    domestic, 68–69; expanded definition
    of, 68–69; Iran's link to, 34–35, 38;
    jihadi, 127; and loss of freedom, 71–72;
    potential targets of, 11; safeguards
    against, 67; training in, 120; and the
    TSA, 64; *see also* war on terror
thermobaric weapons, 51, 119
3M, 79
Tillman, Patrick "Pat," 52–53, 62
Torrijos, Omar, 121
torture, 50, 114–120
Tower 22, 141
Transformers franchise, 28
Transportation Security Administration
    (TSA), 24, 63–64
treason, 106–107
Trump, Donald, 53, 78, 80, 86
Trump administration, 103
truth, 17, 22–23, 110
Tuareg rebels, 127
Tunisia, 127
Twitter (now X), 40, 42–43, 53, 108–109

U
Ukraine, 3–4, 5, 97, 108, 128, 141
United Nations: and the Iraq War, 36;
    Office of the High Commissioner for
    Human Rights, 8; report on deaths in
    Afghanistan, 51; Security Council, 131

United Nations Convention Against
   Torture (UNCAT), 114
United States Central Command
   (CENTCOM), 42
United States Military Academy, 26
Universal Declaration of Human Rights
   (1948), 114
urban guerrilla warfare, 120
USA PATRIOT Act, 68–69
U.S. Customs and Border Patrol, 26
U.S. military: Army Rangers, 52; cost of
   maintaining, 85–95; disability benefits,
   57–58; elevated status of, 56; healthcare
   for veterans, 57; in Japan, Germany,
   and South Korea, 85; in Jordan,
   141; as military-economic fascism,
   82; and professional sports, 25–26;
   recruitment efforts, 58–59; recruitment
   of noncitizens, 59; recruitment of
   women, 59–60; rogue members of, 50;
   social media accounts, 42; support for,
   53; *see also* casualties of war; military
   intervention; Pentagon

V
ventilators, 79
Veteran Affairs (VA) medical system, 57
Veterans of Foreign Wars, 130
Vieques Island, 69
Vietnam War: financing for, 91–92;
   Kissinger on, 129; lies about, 102;
   protests against, 67, 105; refusal to serve
   in, 104; use of torture during, 117
Viola, Roberto, 121
violence, gender-based, 122; *see also* torture;
   war

W
war: abolition of, 144; "just," 142; *see also*
   casualties of war
war bonds, 91
war crimes, 50, 97, 102, 121
War Industries Board (WIB), 75–76
war on drugs, 5
war on terror: civilian casualties, 9, 49–50;
   costs and consequences of, 141; and

the Espionage Act, 103; and the film
   industry, 27; funding for, 92; protecting
   democracy, 5, 11; targeted killings, 70;
   and the use of torture, 117
War Powers Acts (First and Second), 76–77
War Revenue Act, 86–87
waterboarding, 116, 118
water cure, 115
Waters, Maxine, 94
Watson Institute for International and
   Public Affairs, 92
Watt, J. J., 53
weapons of mass destruction (WMDs),
   34–36, 37, 38
Western Hemisphere institute for Security
   Cooperation (WHINSEC), 120
West Point, 26
"whistleblower" documents, 22
Whistleblower Protection Act (1989), 103
whistleblowers, 22, 101–103, 106
WikiLeaks, 108
Wilson, Woodrow, 19, 20, 75, 89, 100
Winner, Reality L., 103
women in the U.S. military, 59–61
World Trade Center, 11
World Trade Organization (WTO), 7
World War I: and the Alien Enemies Act,
   100; public information during, 19, 32;
   War Industries Board (WIB), 75–76;
   and the War Revenue Act, 86–87
World War II: Defense Production Act
   (DPA), 77–78; and the film industry,
   26–27; First and Second War Powers
   Acts, 76–77; Japanese internment,
   65–66; Pearl Harbor, 65; public
   information during, 19, 32; rationing
   during, 74; taxation during, 87–88;
   women in the military, 59
Writers' War Board (WWB), 20, 32

X
X (formerly Twitter), 40–43, 53, 108–109

Z
Zaire (Democratic Republic of the Congo),
   112

About the Authors

Christopher J. Coyne is a senior fellow with the Independent Institute and coeditor of *The Independent Review*, professor of economics at George Mason University, associate director of the F. A. Hayek Program for Advanced Study in Philosophy, Politics, and Economics at the Mercatus Center, and coeditor of the *Review of Austrian Economics*. He received his PhD in economics from George Mason University.

He is the author or coauthor of *The Political Economy of Terrorism, Counterterrorism, and the War on Terror; In Search of Monsters to Destroy: The Folly of American Empire and the Paths to Peace; Manufacturing Militarism: U.S. Government Propaganda in the War on Terror; The Economics of Conflict and Peace: History and Applications; Defense, Peace, and War Economics; Tyranny Comes Home: The Domestic Fate of U.S. Militarism; The Essential Austrian Economics; Doing Bad by Doing Good: Why Humanitarian Action Fails; Media, Development, and Institutional Change; Context Matters: Entrepreneurship and Institutions;* and *After War: The Political Economy of Exporting Democracy.*

Professor Coyne is also a contributing author in more than 30 volumes and 150 articles and reviews in scholarly journals, and the editor or coeditor of 11 books. His popular articles have appeared in such publications as *The Hill, Daily News,* the *Boston Review, Fraser Forum,* the *Detroit News,* the *Cedar Rapids Gazette,* the *La-*

fayette Journal & Courier, the *Herald Times Reporter*, the *Des Moines Register*, the *Green Bay Press Gazette*, the *Shreveport Times*, and the *Muscatine Journal*.

Abigail R. Hall is a senior fellow with the Independent Institute, associate professor of economics at the University of Tampa, an affiliated scholar with the Mercatus Center at George Mason University, a Non-Resident Fellow with Defense Priorities, and an affiliated scholar with the Foundation for Economic Education. She is the book review editor of the *Review of Austrian Economics*. She received her PhD in economics from George Mason University.

She is the author or coauthor of *The Political Economy of Terrorism, Counterterrorism, and the War on Terror*; *Manufacturing Militarism: U.S. Government Propaganda in the War on Terror*; and *Tyranny Comes Home: The Domestic Fate of U.S. Militarism*.

Professor Hall is also a contributing author in seven volumes and more than 35 articles and reviews in scholarly journals, and the editor or coeditor of two books. Her popular articles have appeared in such publications as *CNBC*, *Forbes*, *Newsweek*, and *USA Today*. She has appeared on outlets such as *Fox Business*, PBS, and CSPAN.

Independent Institute Studies in Political Economy

THE ACADEMY IN CRISIS
edited by John W. Sommer

AGAINST LEVIATHAN
by Robert Higgs

AMERICAN HEALTH CARE
edited by Roger D. Feldman

AMERICAN SURVEILLANCE
by Anthony Gregory

ANARCHY AND THE LAW
edited by Edward P. Stringham

ANTITRUST AND MONOPOLY
by D. T. Armentano

AQUANOMICS
edited by B. Delworth Gardner & Randy T. Simmons

ARMS, POLITICS, AND THE ECONOMY
edited by Robert Higgs

A BETTER CHOICE
by John C. Goodman

BEYOND POLITICS
by Randy T Simmons

BOOM AND BUST BANKING
edited by David Beckworth

CALIFORNIA DREAMING
by Lawrence J. McQuillan

CAN TEACHERS OWN THEIR OWN SCHOOLS?
by Richard K. Vedder

THE CHALLENGE OF LIBERTY
edited by Robert Higgs & Carl P. Close

THE CHE GUEVARA MYTH AND THE FUTURE OF LIBERTY
by Alvaro Vargas Llosa

CHINA'S GREAT MIGRATION
by Bradley M. Gardner

CHOICE
by Robert P. Murphy

THE CIVILIAN AND THE MILITARY
by Arthur A. Ekirch, Jr.

CRISIS AND LEVIATHAN, 25TH ANNIVERSARY EDITION
by Robert Higgs

CROSSROADS FOR LIBERTY
by William J. Watkins, Jr.

CUTTING GREEN TAPE
edited by Richard L. Stroup & Roger E. Meiners

THE DECLINE OF AMERICAN LIBERALISM
by Arthur A. Ekirch, Jr.

DELUSIONS OF POWER
by Robert Higgs

DEPRESSION, WAR, AND COLD WAR
by Robert Higgs

THE DIVERSITY MYTH
by David O. Sacks & Peter A. Thiel

DRUG WAR CRIMES
by Jeffrey A. Miron

ELECTRIC CHOICES
edited by Andrew N. Kleit

ELEVEN PRESIDENTS
by Ivan Eland

THE EMPIRE HAS NO CLOTHES
by Ivan Eland

THE ENTERPRISE OF LAW
by Bruce L. Benson

ENTREPRENEURIAL ECONOMICS
edited by Alexander Tabarrok

FAILURE
by Vicki E. Alger

FINANCING FAILURE
by Vern McKinley

THE FOUNDERS' SECOND AMENDMENT
by Stephen P. Halbrook

FUTURE
edited by Robert M. Whaples, Christopher J. Coyne & Michael C. Munger

GLOBAL CROSSINGS
by Alvaro Vargas Llosa

GOOD MONEY
by George Selgin

GUN CONTROL IN NAZI-OCCUPIED FRANCE
by Stephen P. Halbrook

GUN CONTROL IN THE THIRD REICH
by Stephen P. Halbrook

HAZARDOUS TO OUR HEALTH?
edited by Robert Higgs

HOT TALK, COLD SCIENCE, 3RD ED.
by S. Fred Singer with David R. Legates & Anthony R. Lupo

HOUSING AMERICA
edited by Randall G. Holcombe & Benjamin Powell

IN ALL FAIRNESS
edited by Robert M. Whaples, Christopher J. Coyne & Michael C. Munger

IN SEARCH OF MONSTERS TO DESTROY
by Christopher J. Coyne

IS SOCIAL JUSTICE JUST?
edited by Robert M. Whaples, Christopher J. Coyne & Michael C. Munger

JUDGE AND JURY
by Eric Helland & Alexander Tabarrok

LESSONS FROM THE POOR
edited by Alvaro Vargas Llosa

LIBERTY FOR LATIN AMERICA
by Alvaro Vargas Llosa

LIBERTY FOR WOMEN
edited by Wendy McElroy

LIBERTY IN PERIL
by Randall G. Holcombe

LIVING ECONOMICS
by Peter J. Boettke

100 Swan Way, Oakland, California 94621-1428, U.S.A.
Telephone: 510-632-1366 • Facsimile: 510-568-6040 • Email: info@independent.org
www.independent.org